Samuel Downing

The Elements of Practical Hydraulics

For the Use of Students in Engineering and Architecture ..

Samuel Downing

The Elements of Practical Hydraulics
For the Use of Students in Engineering and Architecture ..

ISBN/EAN: 9783743686953

Printed in Europe, USA, Canada, Australia, Japan

Cover: Foto ©berggeist007 / pixelio.de

More available books at **www.hansebooks.com**

ELEMENTS

OF

PRACTICAL HYDRAULICS,

FOR THE USE OF

STUDENTS IN ENGINEERING AND ARCHITECTURE.

PART I.

WITH NUMEROUS WOODCUTS.

BY

SAMUEL DOWNING, LL.D.,

PROFESSOR OF CIVIL ENGINEERING IN THE UNIVERSITY OF DUBLIN;
HON. MEMBER OF THE INSTITUTE OF MECHANICAL ENGINEERS;
ASSOCIATE INSTITUTION OF CIVIL ENGINEERS.

Third Edition, Revised and Enlarged.

LONDON:

LONGMANS, GREEN, AND CO.

1875.

DUBLIN .
PRINTED AT THE UNIVERSITY PRESS,
BY M. H. GILL.

INDEX.

a

INTRODUCTION.

THE science of Hydraulics has for its object the knowledge of the phenomena of fluids in motion, and of the laws which regulate the production of these phenomena.

Applied as an art, its object is to render this know· ledge available in the designs of the civil engineer, as in the determination of the dimensions of pipes for conveying water, gas, or air, and also in works for the collecting, conveying, and distributing the necessary supply of water, for mill-power, or for the summit-levels of canals; or for the supply of cities; and, generally, of all such works as depend for their suitable construction and proportions upon the result of calculations requiring a knowledge of the pressure and motion of fluids.

2. Fluids are defined to be bodies whose particles, by reason of their extreme mobility, yield to every the least force; they have, however, a certain degree of adherence

B

to the lb. ; giving about 36 cubic feet to a ton, or 6 tons to a cubic fathom.

By a like approximation we have 6.25 imperial gallons to the cubic foot. These numbers give rise to many convenient practical rules, which are given in the "Practical Examples," for Chap. I.

The Imperial Bushel, which is the dry measure of capacity, is equal to eight gallons, or 1.29 cubic ft.

Throughout this work, the only units made use of are the foot and the cubic foot. We have in English works on Hydraulics a great variety of units : for volume, the gallon, the cubic foot, the ton, the cubic yard, and the hogshead ; for length, the fathom, the yard, foot, and inch, which, coupled with the absence of decimal subdivisions in our weights and measures, is always perplexing to the reader.

As soon as the student has become familiar with the value of the inches in a foot expressed decimally, it is hoped that this arrangement will be found useful. Of the eleven decimal fractions for the inches in a foot, five are well known, namely, those for $\frac{1}{4}$, $\frac{1}{3}$, $\frac{1}{2}$, $\frac{2}{3}$, $\frac{3}{4}$, and the rest may be readily remembered. It will be observed, also, that the eighth of an inch is very nearly 0.01 ft., and every other eighth has, in the place of hundredths, a corresponding figure, thus—

$\frac{2}{8} = 0.0208$ ft., $\frac{3}{8} = 0.0312$ ft., $\frac{4}{8} = 0.0416$ ft., $\frac{5}{8} = 0.0521$ ft., $\frac{6}{8} = 0.0625$ ft., $\frac{7}{8} = 0.0729$ ft.

Table showing the Decimal Values of the Inch.

Inches.	Fractions of a Foot.		Inches.	Fractions of a Foot.	
1	$\frac{1}{12}$,	0.0833	7	$\frac{7}{12}$,	0.5833
2	$\frac{1}{6}$,	0.1666	8	$\frac{1}{3}$,	0.6666
3	$\frac{1}{4}$,	0.2500	9	$\frac{3}{4}$,	0.7500
4	$\frac{1}{3}$,	0.3333	10	$\frac{4}{5}$,	0.8333
5	$\frac{5}{12}$,	0.4166	11	$\frac{11}{12}$,	0.9166
6	$\frac{1}{2}$,	0.5000	12	$\frac{12}{12}$,	1.0000

The measure of the force of gravity is the velocity acquired in one second by a body falling freely from a state of rest, and is equal to 32.1948 feet per second, and always denoted by the letter *g*.

5. So many French works on Hydraulics, of great value, have been composed, that a notice of their weights and measures may here be useful.

The Mètre, adopted in France in 1798, as the unit of lineal measures, is supposed to be equal to the one ten millionth part of the quadrant of a Meridian of the earth ; the accuracy of this is not, however, essential to the value of the system ; expressed in English measures it is equal to 39.37079 inches, or 3.280899 ft. ; which, in practice, may be taken, approximately, as 39.37 inches, and 3.281 feet. It is multiplied, decimally, into the Decameter, the Hectometer, and Kilometer, and is subdivided, decimally, into the Decimeter, the Centimeter, and the Millimeter; the Greek word being affixed for multiplication, and the Latin for division by ten.

The unit of weight is the Gramme, which is equal to

the weight of a cube of distilled water, at a temperature of 44° centigrade, above zero (supposed to be its maximum density), and *in vacuo;* the side of the cube being one centimeter in length. As the decimeter is equal to ten times the centimeter, its cube will be 1000 times the cubic centimeter; the Kilogram therefore (1000 grammes) is the weight of a cubic decimeter, or liter, of distilled water at the above temperature.

It is equal to 2.20485 lbs avoirdupois, hence 1000 kilos are nearly one ton or 36 cubic feet of water.

The measures of length, area, capacity, and weight, are in this sytem mutually connected; it is not so in the English weights and measures; the side of a cube containing one gallon cannot be expressed by any whole number of inches, or any other lineal measure, as the foot, &c.; it is a little greater than 6.5 inches. Hence the long columns of specific gravities, which are not needed in the metric system, as the weight of any body expressed in Kilograms, whose volume is stated in cubic meters, is also its specific gravity, or ratio of its weight to the weight of an equal bulk of water.

ELEMENTS OF HYDRAULICS.

CHAPTER I.

ON THE FLOW, THROUGH AN ORIFICE, OF WATER CONTAINED IN A VESSEL.

THE vessel from whence water issues through an orifice may be, first, maintained at a constant height of surface; or, secondly, it may receive no supply, and, consequently, be gradually exhausted; or, thirdly, the orifice, instead of discharging freely into the air, may do so into another reservoir, under more or less resisting counter-pressure; and hence *three divisions* of this part of the subject. The second division also includes the cases in which the level of the surface gradually rises or falls, from the supply being greater or less than the discharge through the orifice.

7. The opening through which the water flows may be placed either in the bottom or in one of the sides of the experimental tank, most generally the latter, in which case the surface of the water in the basin should be above the upper edge of the orifice: this orifice is opened either in a thin plate,—that is to say, in a plate whose thickness does not exceed half the diameter of the orifice, if circular, or smallest dimension, if rectangular;—or else it is furnished with an adjutage, or short tube,

sometimes cylindrical, sometimes conical, converging towards an external point, less often diverging. An orifice placed in a very thick plate would evidently be equivalent to one of the same diameter if placed in a thin plate, with an adjutage attached.

We may also have the surface of the fluid below the upper edge of the orifice; that part of the border or circumference is then without influence on the discharge, and very frequently it is not applied; the opening, unlimited on its upper part, is then called an overfall or weir. The laws of the flow of water in this second case offer some peculiarities, and form the subject of a separate investigation. When the surface reaches to a very small height only above the opening, we also have special circumstances: this case is intermediate between the two others first mentioned.

Before entering upon them it is necessary to state briefly the general principles of the flow of water, and the modifications which the " contraction" of the fluid vein suffers in passing through the various orifices to be noticed. The vertical distance of the surface of the fluid above the centre of gravity of the orifice is called the *charge* of the water upon the orifice, or the *head* under which the flow takes place. This point is not the true depth at which the mean velocity is found, but may, in most cases, without any sensible error be taken to represent it; the exact determination of it will be found in a future page.

8. *Velocity of Water flowing from an Orifice.*—Let a vessel X, maintained constantly full of water up to the level AB, have upon the horizontal faces CD and EF the open orifices M and N; the fluid will issue in vertical jets, which will rise almost to the level of the water AK ; they would rise fully up to it but for the resistance of the air. Now, by the first principles of Dynamics, in order

that a body impelled in a vertical direction should reach
to any height, it is necessary that at the point of depar-
ture it should have had a velocity equal to that which
it would have acquired in falling freely from that height;
consequently, the particles of the fluid must have had a

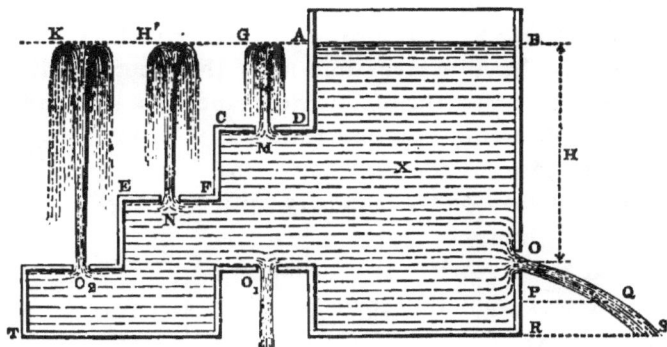

Fig. 1.

velocity nearly equal to that due to the charge—that is,
to the height of the surface of the water above the orifices,—the only supposition in the application of the
principle to the flow of water being that the particles of
the fluid are perfectly independent of each other after
they leave the orifice.

So also, if upon a vertical face BR an orifice be
placed, the centre of which is at O, we shall see further
on that, from the respective values of the lines OP and PQ,
the fluid must have issued from O with a velocity due to
the height OB. It would issue with a velocity due to BR,
if the orifice had been opened in the bottom RT of the
vessel ; and the velocity is the same in O, in O_1, and O_2,
the directions being different, but the charge the same.
This truth holds good for different orifices, whatever may
be the ratio of the area of the orifice to the horizontal
section of the water in the vessel, provided that the level
of the water is kept at the same constant height, and

tranquil ; which last, however, cannot be attained, if the orifice be too large in proportion to that surface—the water of the supply, in that case, producing disturbing movements in the reservoir.

A second method of determining the velocity is by measuring the ordinates of the curve of the path of a jet of water issuing from an orifice in the vertical side of a cistern.   To have a clear notion of this method, it is necessary to state the following principles :—When a body is projected in any direction AY, with a certain uniform velocity, the combined action of this velocity with the force of gravity causes it to describe a curved path, AMB.   By measuring the absissa x and the corresponding ordinate y, we can compute the height from which a body must fall vertically by the action of gravity to acquire that velocity, and, lastly, comparing the height so computed with the actual

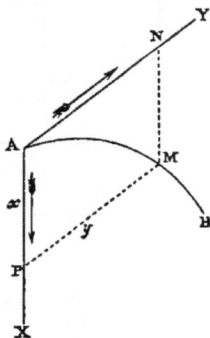

Fig. 2.

height of the surface of the water above the centre of the orifice, they are found to be very nearly equal, and thus we have another proof that water issues from an orifice with a velocity equal to that it would acquire in falling from a height equal to the " head" or charge.

We do not for this purpose require any of the properties of the curve of the jet.   If the velocity, and consequently the resistance of the air, be not very great, the curve is a parabola.   The demonstration of this will appear from the computation of the quantity more immediately sought for, which results in the equation of that curve, the parameter being equal to four times the height due to the velocity of projection.

Let v be the velocity with which the body is sent forth in the direction of AY, and t the time spent in reaching the point N; then, since the velocity in the direction AN is uniform, $AN = v \times t$; on the other hand, if the body had been solely under the action of the accelerating force of gravity, it would have descended from A to a point P, during that same interval t, such that we should have $AP = \frac{1}{2}gt^2$. If we complete the parallelogram APMN, the point M will have been reached under the joint action of these movements in the same time t in which the point P was attained under the sole accelerating force; and it will have, therefore, traversed the arc of the curve, whose abscissa will be AP, and ordinate MP, parallel to the axis AY. Let $x = AP$ and $y = MP$, we have therefore from the laws of gravity

(a) . . . . $x = \dfrac{gt^2}{2}$

and from the uniform velocity in the direction AY

(b) . . . . $y = vt.$

From (b) we have by division, $t = \dfrac{y}{v}$, and squaring $t^2 = \dfrac{y^2}{v^2}$; substituting this value of t^2 in (a), we have $\dfrac{gy^2}{2v^2} = x$, or

(c) . . . . $y^2 = \dfrac{2v^2x}{g} = \dfrac{4v^2x}{2g},$

and putting h for the height due to the velocity v, and remembering that $\dfrac{v^2}{2g} = h$, we have

(d) . . . . $\dfrac{y^2}{4x} = h,$ and also

(e) . . . . $\dfrac{y^2}{2x} \cdot g = v^2.$

This truth, which has been proved for any body in general, holds good also for a jet of water issuing from an orifice. If this orifice be opened in a vertical plate, the axis of projection being horizontal, the ordinates—that is, the distances of the different points of the jet from the vertical let down from the centre of the orifice —will be horizontal; and if

Fig. 3.

through any point C of this vertical we draw a horizontal plane, then, according to the theorem, the square of the distance CD—called the range of the jet—taken in this plane (or generally of any distance MP), divided by four times the corresponding fall, AC, will give the height due to the velocity of exit, as from (*d*) we have $h = \dfrac{y^2}{4x}$.

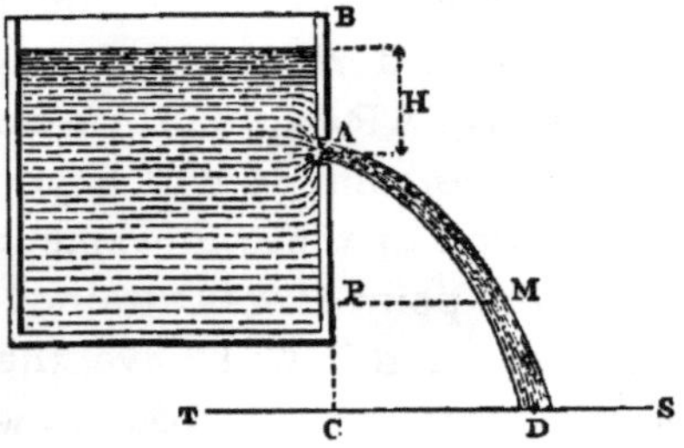

And the permanent form that the jet of water assumes being identical with the path of any single particle acted on by the same forces, we are enabled to use it as a mode of measuring the velocity of the water at its issue from the orifice. A vertical rod divided into any scale of equal parts, and firmly fixed, having its zero at the centre of the orifice, has applied to it at right angles another rod similarly divided, and having a stock like a T square, so as to slide up or down the vertical fixed rod, and its zero being in the vertical let fall from the mouth of the orifice, we can then measure any ordinate PM and corresponding abscissa AP. Hence, by measuring *y* and *x*, we can calculate *h*, the height due to the velocity of exit, from the formula just given. And comparing *h* so found with H, the charge, we find them very nearly coincident, as in the following Table of experiments by Bossut :—

Curve of the Jet.		Height due to Velocity of Exit Calculated.	Charge Measured.	Differences.
Abscissa = x.	Ordinate = y.			
Feet.	Feet.	Feet.	Feet.	Feet.
20.598	24.698	7.404	7.511	0.107
15.284	27.716	12.564	12.890	0.326
4.624	20.500	22.720	23.583	0.863

Thus $\dfrac{24.698^2}{4 \times 20.598} = \dfrac{610.1}{82.4} = 7.404$, and so of the other numbers in the third column.

The difference between the third and fourth columns inereases with the charge, and we should expect it to be so, since the cause of this difference—the resistance of the air—increases as the square of the velocity, and, consequently, nearly as the charge. Had it not been for this, the difference would have been very nearly equal to zero, and the velocity at the section of contraction, as mentioned above, is truly stated as equal to that due to the charge. This general proposition may consequently be laid down :—" Water flowing through an orifice in a vertical thin plate issues with a velocity, $q, p,$ equal to that due to the charge."

Thus the velocity acquired by a body falling freely by the force of gravitation from the height H, is equal to that of the fluid as it issues from the orifice with that height for the charge; that is—

$$V = \sqrt{2g\mathrm{H}},$$

called after Toricelli, its discoverer, the Torricellian theorem, in which H is the " charge," measured from the surface down to the centre of gravity of the orifice, and *g*, the dynamical measure of the force of gravitation, being the rate, or number of feet per second, with which a body falling freely is moving at the end of the first second.

9. The following Table exhibits the results of experiments by Castel, D'Aubuisson, Bossut, Poncelet, &c., also proving that the velocity of issue is proportional to the square root of the charge.

It will be observed that the charges vary from 1 to 200 and more, and the sections of the orifices from 1 to 500, and yet in all cases the velocities have followed the ratio of the square roots of the charges, minute discrepancies, sometimes giving too great a number, and sometimes too small, being inseparable from experiments of this nature.

The actual object of measurement in the experiments was the *quantity* discharged in a given time; but it is evident that, with the *same orifice*, the discharge is exactly proportional to the velocity with which the fluid issues, and, therefore, that column in the Table which expresses the gauged discharges, reference being made to some one discharge as a unit, also expresses the velocities.

Thus, for example, take the 1st and 4th lines with "square orifice." The discharge of the former into a cubical vessel was 229 cubic feet in 40 seconds, and of the latter 258.5 cubic feet in 25 seconds; reducing both to the quantity for one second, we have 5.725 and 10.34 cubic feet respectively, dividing each by 5.725 their ratio is 1 to 1.806. The square roots of the charges are 1.1454 and 2.0652, dividing both by the first their ratio is 1 to 1.803, as in the two last columns of the Table.

TABLE *showing that the Velocities are proportional to the square roots of the Charges.*

Diameter of the Orifice.	Charge above the Orifice.	Series of	
		Square Roots of the Charges.	Discharges or Velocities.
Feet. 0.0328	Feet. 0.085	1.000	1.000
	0.098	1.074	1.064
	0.131	1.241	1.244
	0.164	1.386	1.393
	0.196	1.519	1.524
0.088	4.265	1.000	1.000
	9.580	1.500	1.497
	12.500	1.713	1.707
0.265	7.677	1.000	1.000
	12.500	1.305	1.301
	22.179	1.738	1.692
0.531	6.922	1.000	1.000
	12.008	1.316	1.315
Square Orifice 0.656 by 0.656	1.312	1.000	1.000
	2.296	1.323	1.330
	3.281	1.581	1.590
	4.265	1.803	1.806
	5.249	2.000	2.000

10. The general principle, that the velocities are as the square roots of the charges, as also the theorem of Torricelli (§ 8) for cases in which it is applicable, extends to every kind of fluids,—to mercury, oils, alcohols, so that the velocity with which each of them issues from an orifice is independent of its particular nature, and of its density, it depends solely on the charge. Experiment demonstrates this, and very simple reasoning suffices to show its truth. Take the case of mercury : the particles situated immediately in front of the orifice, and in which it is necessary to create a certain velocity, are, it is true, fourteen times more dense than those of water, and they consequently oppose to motion a resistance fourteen

times greater than it would do ; but the mass also which
presses upon these particles, and produces the velocity
of exit, the charge being the same, is greater in the same
proportion, and therefore gives a motive force fourteen
times greater.   Thus a compensation exists, and the
velocity impressed remains the same ; and, in like manner, it may be proved for a fluid lighter than water.

11. The proposition that has now been laid down
with respect to the velocity of water issuing through
an orifice is equally true in cases when the discharge
takes place *in vacuo*, the velocity is always the same,
with the same head, whatever be the pressure upon the
free surface of the water in the vessel, provided the jet
of water at its exit from the orifice be subject to an equal
exterior pressure.   But if the pressures on these two
surfaces be not equal, the velocity will be very different
from that due to H.

If in the first place the pressure per square inch
against the orifice at A be greater than that upon the
free surface of the water BC (in the woodcut, Fig. 4),
then the excess of the former above that on the free
surface must be less than that of a column of the fluid
whose height is the vertical distance of the orifice A,
below the surface BC, for if they were equal, it is evident
there could be no discharge.

Let us, then, take an horizontal plane DE, below the
plane BC, at such distance from it that the weight of a
column of the water contained between the two planes,
and whose base is the unit of surface, may be equal to
the excess of pressure at A, of which we are speaking.
The pressure, then, which exists upon any point in
the plane DE will be equal to that upon any point in
BC, *plus* the supposed excess of pressure against the
exterior of the orifice ; and, therefore, the pressure
upon the plane DE, will be the same as that against the

orifice at A. The liquid below the plane DE is then in the same condition as if that contained between BC and and DE were removed, and the free surface and exterior of the orifice were under equal pressures; and thus the formula

$$V = \sqrt{2g(H - h_1)},$$

will represent the velocity; h_1 denoting the depth of the plane DE below BC.

The water in the vessel being supposed to have but a slight degree of motion, on account of the relatively small area of the orifice to the surface BC, which is understood to subsist; and therefore we may assume the pressures to be transmitted as if the water was in equilibrium.

12. If, secondly, the exterior pressure on A were less than that upon the surface BC, we may conceive the excess of pressure on BC to be produced by a liquid of the same specific gravity as that in the vessel, applied above BC and terminating in a free surface D′E′, situated at such height that the vertical distance represents, as before, the column of the liquid, having for its base one square inch, or other unit of surface, whose pressure is equal to the excess of the pressure on BC above that against the orifice A. The flow, then, will take place with the same velocity as if the free surface of the liquid, instead of being in the plane BC, and supporting this excess of pressure, were at D′E′, and supported the same pressure as the orifice at A; the formula will therefore be

$$V = \sqrt{2g(H + h_2)},$$

in which h_2 is equal to the vertical distance of D′E′,

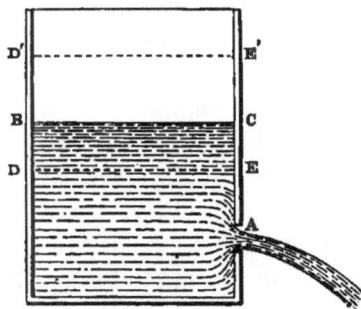

Fig. 4.

above BC. We see thus that a diminution or augmentation of the pressure upon the free surface of the liquid in the vessel, without any change in that against the orifice at A, causes a corresponding diminution or augmentation in the velocity of the issuing fluid, and, on the contrary, that a diminution or augmentation of the pressure against the orifice, without any change in that upon the free surface, causes a corresponding augmentation or diminution in this velocity.

The self-acting contrivance (of James Watt) for supplying the feed-water to low-pressure boilers comes under the first case. The pressure being supposed 5lbs. per inch above the atmosphere, it is required to place the cistern of the supply so high, that on the opening of the valve *a*, by the float *b* descending below the proper level, the water may enter against the pressure of the steam. Now, as the cubic foot of water weighs 62.5 lbs., a column 1 foot high and 1

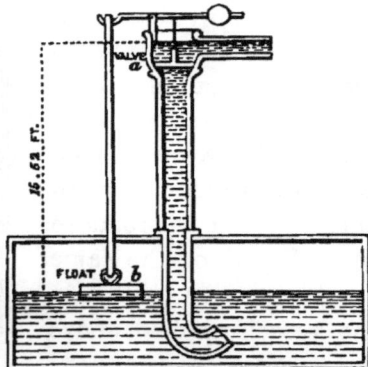

Fig. 3.

square inch base weighs $\frac{62.5}{144}$ = 0.434 lbs., and, therefore, the height of the column of water to balance any given pressure expressed in pounds per square inch is found by dividing that number by 0.434 in this case, 5 ÷ 0.434 = 11.52 feet: this gives exact equilibrium; the additional head, in order that it may enter with due rapidity (from 2 to 4 feet per second generally), will depend upon the rate of evaporation of the boiler and the area of the supply pipe. It is evident that this mode of supply is not convenient in high-pressure boilers; for suppose the pressure to be 50 lbs. per inch, then the height to produce equilibrium will be 115.2 feet. The pressure in a hydraulic press is frequently 3 tons per square inch, equal to (3 × 2240 =) 6720 lbs., and 6720 ÷ 0.434 = 15484 feet.

If, instead of a free surface in the cistern, we had supposed a solid piston or plunger to press on the enclosed water, the head should in like manner be calculated, by turning the pressure per square inch on the piston into vertical feet of water.

The condenser of a low-pressure steam-engine offers an example of the second case; for, let us suppose a vacuum of 25 inches of mercury to be maintained, and that the head of water in the cistern supplying the jet of cold water which effects the condensation, were 2 feet above the point at which it enters this partial vacuum, then the actual head producing the flow is 2 + 28.25 = 30.25 feet, for, pure mercury being 13.56 times heavier than water, we have the height of a column of water which would balance that of 25 inches of mercury equal to 25 × 13.56 = 339 inches, or 28.25 feet.

13. Having thus established the law of the velocity of a fluid issuing from an orifice, let us proceed to apply it to the determination of its discharge, which is defined to be the volume of the fluid which escapes in the unit of time, that is, one second.

If the mean velocity of all the particles was that due to the "charge" H, then this velocity, which is called the theoretic velocity, would be $\sqrt{2gH}$; and if at the same time the particles issued from all points of the orifice in parallel threads, it is evident that the volume of water flowing out in a second would be equal to the volume of a prism which would have the orifice for its base, and that velocity for its length; and, calling S the area or section of the orifice, the volume of water, or of the prism, would be—

$$S \times V = S\sqrt{2gH}.$$

This is the theoretic discharge.

14. But the actual discharge is always less than this. In order to have an exact idea of the phenomena, let us consider the fluid vein a short distance after its issue from the orifice, and let us suppose it cut by a plane perpendicular to its direction. It is manifest that the discharge will be equal to the product of the section by the mean velocity of all the several threads at the moment they intersect the plane of the section. If this section was equal to that of the orifice, and if this velocity was that due to the charge, then the actual discharge would also be equal to the theoretic discharge. But whether from the section of the vein being considerably less than that of the orifice,—as in the flow through orifices in a thin plate, or from the velocity being considerably less than that due to the charge, as in cylindrical adjutages ; or, again, from a diminution in both the section and the velocity, as in certain conical adjutages,—it always results, that the actual discharge is in every case less than the theoretic, and, in order to reduce this last to the former, it is necessary to multiply it by some fraction.   Let m represent this fraction, and Q the actual discharge in one second, we shall then have—

$$Q = m\,S\,\sqrt{2gH}.$$

And representing the volume of water flowing off in the time T seconds by Q' we shall have—

$$Q' = m\,ST\,\sqrt{2gH}.$$

Whether the diminution in the discharge arises from a diminution of the section, or of the velocity, it is always a consequence of the contraction which the fluid vein suffers in passing through the orifice, and thus the multiplier m, or " coefficient for the reduction of the theoretic to the actual discharge," is generally called the " *coefficient of contraction*," and is taken to represent the aggregate effect of all circumstances tending to diminish the discharge. Its accurate determination is of the greatest importance ; upon the degree of exactness with which it is ascertained depends that of the results we obtain when we would apply to practice formulæ upon the flow of water. We shall now proceed to give the results of experiments on the value of the symbol *m*, making some preliminary statements upon the cause of the " contraction," and the nature of its effects ; and also upon the form of the fluid vein—the orifice being circular—its relative dimensions, and the effect of the form upon the discharge.

15. *Cause of the Contraction.*—If we take a glass vessel in the side of which is an orifice through which the water flows, and render visible the movement of the molecules of the water in the vessel by disseminating through it a substance of equal specific gravity, and very minute, or by producing within the water some light chemical precipitation, such as occurs when we let fall a few drops of nitrate of silver in water slightly saline,—we then see at a small distance from the orifice, as, for instance, about an inch, when its diameter is three-eighths of an inch—the fluid molecules converge from all parts towards the orifice, describing curved lines, and, finally, as if approaching a centre of attraction, issue forth with a rapidly increasing motion.

The convergence of the directions that they had within the vessel at the moment of their arrival at the orifice still continues for a short distance after they have passed out, so that we can plainly see the fluid vein gradually diminish, and become contracted up to the place where the particles, from the effect of their mutual action, and of the motions impressed upon them, take directions, either parallel to each other, or in some other lines. The vein thus forms a species of truncated pyramid or cone, whose larger base is the orifice, and smaller the section of the fluid at its place of greatest contraction, a section which is often called the "section of contraction." This figure, and all the phenomena of contraction, are thus a consequence of the convergence of the several threads of water when they arrive at the orifice.

16. *Effects of the Contraction.*—When the orifice is in a thin plate, the contraction is completely external to the reservoir; it is thus clearly visible, can be, and, in fact, it has been, measured, as we shall mention directly. When the orifice is circular, the fluid vein, after having reached the minimum section, continues of the same transverse area, and is thus cylindrical in form, having a velocity very nearly equal to that due to the charge. The discharge will, therefore, be the product of this section by the velocity, so that the effect of the contraction is limited to the reduction of the value of the section which enters into the expression for the value of the discharge. The flow will take place as if the actual orifice had been replaced by another whose diameter was equal to the "section of contraction," but in which supposed orifice no true contraction took place.

17. *Form and Dimensions of the Contracted Vein of the Fluid.*—Let us next examine the form that the contraction gives to the fluid vein issuing from an orifice, in the simple case of a circular orifice in a thin plate, truly plane. Everything being symmetrical around the different points of the orifice, the direction as well as the velocity of the molecules, the contracted vein ought also to be of a symmetrical form, and, consequently, a solid of revolution—a conoidal figure. It is actually so according to the observations that have

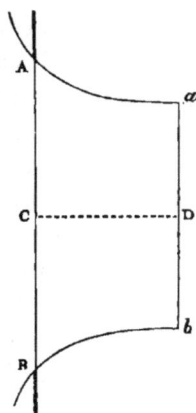

Fig. 6.

been made, and

which the figure AB*ab* represents. Beyond *ab* the contraction ceases, and the vein continues sensibly cylindrical for a certain length, until the resistance of the air and other causes entirely destroy this form.

The earlier measurements that have been made give to the three principal dimensions AB, *ab*, and CD, the ratio of the numbers 1.00, 0.79, and 0.39. The length of the contracted vein would thus be about half the diameter of the smaller section, and 0.39 of the larger, that is, of the orifice.

18. Michelotti, from a mean of more recent experiments on a large scale, has adopted 1.00, 0.787, 0.498 : these D'Aubuisson follows. The ratio of the diameters AB and *ab* being thus 1 to 0.787, that of the sections is 1 to $(0.787^2 =)$ 0.619, that, namely, of the squares of the former numbers; thus, if *s* be the "contracted section," and *S* that of the orifice, we shall have—

$$s = 0.619\ S,$$

and, consequently, the discharge in one second will be

$$s \sqrt{2g\mathrm{H}},\ \text{or } 0.619\ S \sqrt{2g\mathrm{H}},$$

so that the value of *m*, or the " coefficient of contraction," as determined by actual measurement, is, at the mean, equal to 0.619, being a little less than that which results from experiments on the gauged discharge.

If the velocity at the passage of the "section of contraction" was exactly that due to the charge, and that the flow took place through an adjutage of the exact form of the contracted vein, and that in the expression for the discharge the area, *s*, of the outer orifice of this adjutage, taken at the extremity, were introduced, then the calculated would be equal to the actual discharge, and the coefficient of the reduction of the one to the other would be equal to unity; and Michelotti, in one of his experiments in which he employed a cycloidal adjutage, has reached 0.984. It is very probable he would have actually reached 1, if this form had more accurately been adapted to that of the fluid vein, and if the resistance of the air had not somewhat retarded the motion.

19. *Flow of Water through an Orifice in a Thin Plate.*— We come now to the more direct determination of the coefficient for reducing the theoretical to the actual discharge. For this purpose it is necessary to gauge with care the volume of water discharged in a given time under a constant charge, from which we deduce the flow in one second, or the actual discharge; and, dividing this by the theoretic discharge for the same head and same orifice, the quotient is the *coefficient* required.

Thus in a cistern with a head of 4.012 feet above the centre of an orifice the diameter of which is 3.185 inches we have a theoretic discharge of 0.8903 cubic feet per second, obtained thus: with the above head we have a velocity of 16.07 feet per second, that is $\sqrt{2 \times 32.2 \times 4.012}$. And the area of the orifice is equal to

and

$$3.185^2 \times 0.7854 = 7.97 \text{ square inches,}$$

$$\frac{7.97}{144} = 0.0554, \text{ its value in square feet.}$$

This, multiplied into the velocity of issue, gives the volume of the prism or cylinder equal to that of the water discharged;

that is (§ 13), $0.0554 \times 16.07 = 0.8903$ cubic feet per second.

But having found by experiment that in $1\frac{1}{2}$ minutes the actual discharge was 49.68 cubic feet, reducing this to its value for one second by dividing by 90, we obtain

$\dfrac{49.68}{90} = 0.552$ cubic feet as the discharge in one second;

hence, dividing the actual by the theoretic discharge, we

find for the *coefficient* $\dfrac{0.552}{0.8903} = 0.620$.

Many hydraulicians have for a long time been engaged in its determination. The following Table, from D'Aubuisson, gives the principal results obtained by experiments up to the present time, and which, having been made under favourable circumstances, are generally received. They include circular, square, and rectangular orifices :—

TABLES *of the Results of Experiments undertaken to determine the " Coefficient of Contraction."*

CIRCULAR ORIFICES.			
Observers.	Diameters.	Charges.	Coefficients.
	Feet.	Feet.	
Mariotte, .	0.0223	5.8712	0.692
Do.	0.0223	25.9120	0.692
Castel, . .	0.0328	2.1320	0.673
Do.	0.0328	1.0168	0.654
Do.	0.0492	0.4526	0.632
Do.	0.0492	0.9840	0.617
Eytelwein, .	0.0856	2.3714	0.618
Bossut, . .	0.0889	4.2640	0.619
Michelotti, .	0.0889	7.3144	0.618
Castel, . .	0.0984	0.5510	0.629
Venturi, . .	0.1345	2.8864	0.622
Bossut, . .	0.1771	12.4968	0.618
Michelotti, .	0.1771	7.2160	0.607
Do.	0.2657	7.3472	0.613
Do.	0.2657	12.4968	0.612
Do.	0.2657	22.1728	0.597 ?
Do.	0.5314	6.9208	0.619
Do.	0.5314	12.0048	0.619

SQUARE ORIFICES.

Observers.	Side of Square.	Charges.	Coefficients.
	Feet.	Feet.	
Castel, . .	0.0032	0.1640	0.655
Bossut, . .	0.0885	12.5000	0.616
Michelotti, .	0.0885	12.5000	0.607
Do.	0.0885	22.4078	0.606
Bossut, . .	0.1771	12.5000	0.618
Michelotti, .	0.1771	7.3472	0.603
Do.	0.1771	12.5624	0.603
Do.	0.1771	22.2384	0.602
Do.	0.2689	7.3489	0.616
Do.	0.2656	12.5624	0.619
Do.	0.2656	22.3700	0.616

RECTANGULAR ORIFICES.

Rectangle.		Charges.	Coefficients.
Height.	Base.		
Feet.	Feet.	Feet.	
0.0301	0.0606	1.0824	0.620
0.0301	0.1213	1.0824	0.620
0.0301	0.2423	1.0824	0.621
0.0301	0.4847	1.0824	0.626

20. The experiments of Michelotti were carried on about three miles from Turin, at an hydraulic establishment constructed for experimental purposes, consisting of a building 26 feet high, supplied with water from the River Dora by a canal of derivation. The internal dimension was a square of 3 feet 2¼ inches; on one of the sides was arranged a series of adjutages at the different depths deemed expedient, and upon the surface of the ground were arranged the different receptacles for the gauging of the actual discharges.

It may be remarked upon the part of the Table given

by Michelotti that the coefficients obtained from the large orifices are higher than the others, and this contrary to the rule that would be deduced from the experiments in general.

The older writers supposed the deficiency in the discharge to arise from a diminution in the velocity, and not from the *vená contractá.* Thus Hutton writes : " The particles entering the orifice in all directions impede one another's motion : from whence it appears that the real velocity is less than that of a single particle only, urged with the same pressure of the superincumbent column of the fluid. And experiments on the discharge show that the velocity must be diminished rather more than a fourth, or such as to make it equal to that of a body falling through half the ' charge' above the orifice."

21. In order to place the subject of the variation in the value of the coefficient, under different circumstances of area and charge, in a clear point of view, the following Table of MM. Poncelet and Lesbros' experiments at Metz, in the Province of Lorraine, in 1826 and 1827, is given. In these experiments the orifices were rectangular, and all of the same breadth—namely, $0^m.20 = 0.656$ feet; the heights were successively 0.656, 0.328, 0.164, 0.098, 0.065, and 0.0328 feet. The charges extended from 0.33 feet to 5.58 feet. With the several orifices they repeated the experiments, taking each of them with 8 or 10 charges, from the smallest, to the highest that the apparatus admitted, calculating the corresponding coefficients. They then took the charges for the abscissæ, and these coefficients for the ordinates of a curve constructed for each orifice, and by its aid they determined the ordinates,—that is, the coefficients intermediate to those directly determined by experiment ; and thus gave a very extended Table, from which the following is taken :—

TABLE *showing the Results of Experiments to determine the Variation in the Value of the Coefficient of Contraction.*

Charge on Centre of Orifice.	HEIGHT OF THE ORIFICES.						Difference of maximum and minimum coefficients.
	Feet. 0.656	Feet. 0.328	Feet. 0.164	Feet. 0.098	Feet. 0.065	Feet. 0.032	
Feet.							
c.032						0.709*	
0.065						0·698	
0.098				0.638	0.660*	0.691	
0.131			0.612	0.640	0·659	0.685	
0.164			0.617	0.640	0.659	0.682	
0.196		0.590	0.622	0·640*	0.658	0.678	
0 262		0.600	0.626	0.639	0.657	0.671	
0 328		0.605	0·628	0.638	0.655	0.667	
0.393	0.572	0.609	0.630	0.637	0.654	0.664	0.092
0.492	0.585	0.611	0.631	0.635	0.653	0.660	0.075
0.656	0.592	0·613	0.634*	0.634	0.650	0.655	0.663
0.984	0.598	0.616	0.632	0.632	0.645	0.650	0.052
1.312	0·600	0.617*	0.631	0.631	0.642	0.647	0.047
1.640	0.602	0.617	0.631	0.630	0.640	0.643	0.041
2 296	0.604	0.616	0.629	0.629	0.637	0.638	0.034
3.281	0.605*	0.615	0.627	0.627	0.632	0.627	0.027
4.264	0.604	0.613	0.623	0.623	0.625	0.621	0.021
5.248	0.602	0.611	9.619	0.619	0.618	0.616	0.017
6.562	0.601	0.607	0.613	0.613	0.613	0.613	0.012
9.843	0.601	0.603	0.606	0.607	0.608	0.609	0.008

The woodcut illustrates this method of interpolation. From the point O, the several charges are laid off on the

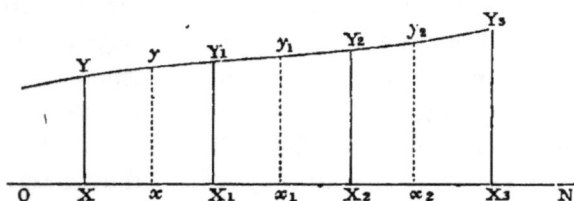

Fig. 7.

line ON, as OX, OX₁, &c., and the corresponding coefficients XY, X₁Y₁, &c.; and the curve being traced through Y, Y₁, Y₂, &c., we can obtain the coefficient proper to any charge as O*x*, by drawing the perpendicular line *xy* terminating in the curve.

22. All the numbers contained in this Table are the several values of the coefficient *m* in the formula $Q = mS\sqrt{2gH}$. But those in each column above the darker type are not the true coefficients for the reduction of the theoretic to the actual value, as will be shown hereafter.

Casting the eye over each column, we may see that the coefficients increase as the charges are greater, but up to a certain point only, although the charge still increases: an asterisk in each column indicates the maximum value. It may also be observed, that the coefficients become more nearly equal in each column as the charges increase,—the bottom line of figures, in which the charge was $3^m = 9.84$ feet, being almost identical in each column.

23. This Table, although constructed from experiments on rectangular orifices, can yet be extended to those of all other forms,—the height of the rectangle, as given in the Table, corresponding to the smaller dimension of the orifice made use of; for it is admitted that the discharge is altogether independent of the figure of the orifice when the area is constant, provided only that this figure has no re-entrant angles.

24. Although these experiments are on a considerable scale, yet there are some cases in practice in which the discharge is twenty or thirty times greater. Such are the sluices in lock-gates on canals of navigation. It is a matter of importance to determine directly the coefficient of discharge for them, and to be able in practice to assign, with some confidence, the coefficient to employ in any particular case, when a direct experiment may not be possible.

TABLE *showing the Value of the " Coefficient of Contraction" in large Sluices: Canal Laquedoc.*

SLUICE.—Width 4.25 ft.		Charge upon the Centre.	Discharge in one sec.	Coefficient.
Area.	Height.			
Square Feet.	Feet.	Feet.	Cubic Feet.	
7.7442	1.804	14.550	145.3056	0.613
6.9928	1.640	6.628	92.6438	0.641
6.9928	1.640	6.245	88.2288	0.629
6.4664	1.508	12.874	138.6302	0.641
6.7237	1.574	13.582	128.7759	0.647
6.7237	1.574	6.392	83.9551	0.616
6.7237	1.574	6.215	79.8580	0.594
6.7172	1.574	6.478	85.2266	0.621
			Mean Coefficient 0.625	

The mean coefficient, 0.625, is rather greater than that found by Poncelet (§ 21), which is readily explained, as the flow of water did not take place in a thin plate, the contraction being suppressed on some parts of the boundary. The wood-work which surrounded the sluice-way was 0.8856 feet thick, and on the sill was even 1.771 ft. Thus, when the sluice was raised but a small height, the contraction nearly ceased on four sides, and the coefficient was considerably increased. For example, when the sluice was raised only 0.393 ft., it gave a coefficient of 0.803 ; when raised 1.51 ft., it was 0.641.

25. *Particular Cases in which the Contraction is suppressed on one or more Sides of the Orifice.*—In all the different cases treated of hitherto, it has been assumed that the fluid arrived at the orifice from all parts equally, but frequently this is not so. For example, Fig. 8, when a rectangular orifice is at the bottom of a vertical plate, and its inferior edge is on the level of the bottom

of the vessel or reservoir, the contraction is then destroyed on that side, the particles of water being compelled to take a direction parallel to the side of the vessel; and, consequently, the discharge is increased. The question arises, therefore, how much will the discharge be augmented by the suppression of the contraction for a certain length of the periphery of the orifice ?

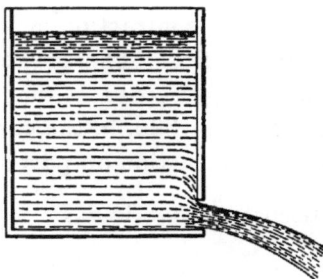

Fig. 8.

The following Table gives the result of experiments instituted with the view of determining this point. The orifice was rectangular, 0.177 feet in base, and 0.089 feet in height. The plates, which were attached, sometimes on one side, sometimes on two or three of the sides, were 0.22 feet long; that is, they advanced this much into the reservoir. The flow was produced by charges from 6.56 feet to 22.56 feet in height :—

TABLE *showing the Increase of the " Coefficient of Contraction" by its Suppression on Part of the Sides.*

Portion of Orifice without Contraction.	Coefficient.	Ratio of Increase.
0	0 .608	1 .000
$\frac{1}{8}$	0 .620	1 .020
$\frac{2}{8}$	0 .637	1 .049
$\frac{3}{8}$	0 .659	1 .085
$\frac{4}{8}$	0 .680	1 .119
$\frac{5}{8}$	0 .692	1 .139

26. In this Table the last column has for its unit the discharge when the orifice is perfectly free : the numbers, therefore, indicate the increase in the coefficients, and,

consequently, in the discharges. The formula deduced by M. Bidone, the experimenter, is $1 + 0.152\frac{n}{p}$, in which n represents the length of the part of the perimeter in which the contraction is suppressed, and p the perimeter of the orifice. The greatest error of this formula being but the $\frac{1}{30}$th part, it may be used for the value of the discharge when, in the case of rectangular orifices, there is no contraction on part of the boundary, and the actual discharge then is $mS\sqrt{2gH}\left(1 + 0.152\frac{n}{p}\right)$.

27. *Orifices in Plates not being true Planes.*—It has been hitherto always supposed that the sides or plates in which the orifices were placed were true planes; they may, however, be of very differently formed surfaces. In order to have a clear idea of the effect which any such alteration produces upon the flow, it is necessary to recall to mind that if the threads of the fluid vein did arrive at the orifice mutually parallel, the actual discharge would be equal to the theoretic, and that it is less than this only by reason of the oblique directions in which

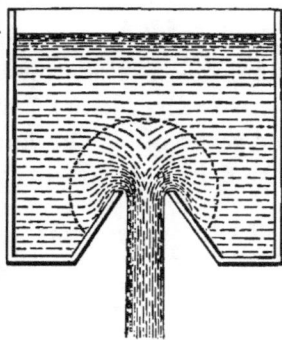

Fig. 9.

they converge, from which necessarily results a destruction of part of the acquired motion at the point of contact with the orifice. If, therefore, we imagine around the orifice a spherical surface of a radius equal to that of the sphere of action of the orifice, and this surface terminated by the sides of the vessel, then it must be intersected on every point, and in directions nearly perpendicular, by the threads of the issuing fluid, as in the woodcut, Fig. 9; and the larger the portion of the complete sphere this surface may be, and the more oblique, or even opposite, to one another, the threads of the fluid arrive at it, then the more the motion is destroyed at the entrance of the orifice, and the less the discharge is found to be. When the sides are developed in one plane, then the supposed surface is a hemisphere, and the coefficient of this particular case is given above. But if they are disposed in the form of a funnel, or, if simply concave, towards the interior of the vessel, then the surface of this sphere is of less extent, and the discharge more considerable—not, however, following exactly the inverse proportion of the spherical surface. If, on the other hand, the side is convex, the discharge is dimin-

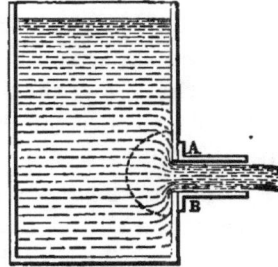

Fig. 10.

ished, and it will be less still in the case represented, Fig. 11. Lastly, it will be at its minimum value if the supposed surface should become an entire sphere; and this would happen if it was possible to carry an orifice into the midst of a mass of the fluid enclosed in the vessel.

28. Borda has succeeded in realizing this case almost completely. He has introduced into a vessel, as shown in Fig. 11, a tube of tin 0.443 feet long and 0.105 feet in diameter, and under a charge of 0.82 feet he has caused the flow to take place so that the effluent water did not touch the tube at all. The actual discharge has been only 0.515 of the theoretic, and from various circumstances Borda was led to think that he might have reduced it to 0.50.

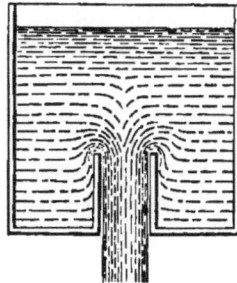

Fig. 11.

The woodcut shows the manner in which the fluid bends around the exterior edge, and enters the tube without touching the internal sides, the thickness being about $\frac{1}{12}$ of an inch, or 0.0069 feet, and the edges cut truly square : thus all that part of the sides within the exterior periphery is, as far as the discharge is concerned, as if totally removed ; and it is this external diameter that should be introduced in all calculations relative to internal adjutages. By taking it, M. Bidone has found, from two experiments in which the effluent fluid did not touch the sides, that the coefficient was nearly 0.50,— that is, the area of the section of contraction was half the area of the orifice taken at the external circumference. Having subsequently surrounded the orifice of the entry of the tube with a border or rim, and having thus reduced it to the condition of being in a plate perfectly plane, although in the centre of the fluid mass, he found the coefficient rise to 0.626. The same result might be obtained by employing a simple tube, but of a thick material.

29. Thus 0.50 and 1.00 will express the limits of the coefficients of contraction,—the limits to which they may approach very nearly, but which they can never actually attain. For orifices in a plate truly plane it does not descend below 0.60, or rise much above 0.70 ; and in ordinary practice it ranges between 0.60 and 0.64. As a mean term, 0·62 is generally taken : so that—

$$Q = mS\sqrt{2gH} = 0.62S\sqrt{2gH} = 4.96S\sqrt{H} ;$$

from whence we have, as an approximate rule for the discharge in cubic feet per second—

$$Q = 5 \times \text{Area} \times \sqrt{H},$$

and per minute, $300 \times \text{Area} \sqrt{H} ;$

and if the orifice be circular of a diameter d, the area is expressed by $d^2 \times 0.7854 = S$, and

$$Q = 3.9 \, d^2 \sqrt{H},$$

or approximately,

$$Q = 4 \times d^2 \times \sqrt{H} \text{ per second,}$$

and $240 \times d^2 \times \sqrt{H}$ per minute, the diameter being expressed in feet. For greater exactness in the coefficient, recourse should, however, be had to the Table, page 28, § 21, that one being chosen which has the nearest identity to the particular case.

30. *Effects on the Discharge when the Fluid has Velocity antecedently.*—If the water contained in the reservoir, instead of being in a state of repose, was moving in the direction of the orifice,—as when the vessel, having but a relatively small section, has a supply of water brought into it, and flowing directly up to the plate or side in which the orifice is opened,—then the particles of the fluid would issue, not only in virtue of the pressure exerted by the fluid mass which is above it, but with the additional velocity that they had when they entered into the sphere of action of the orifice; we must, therefore, add to the actual charge measuring the pressure a new term, which will be the height due to this supposed velocity of arrival. Thus, if u represent this velocity, we shall have (since $\dfrac{u^2}{2g}$ is the height producing the velocity u) the expression—

$$Q = mS\sqrt{2g\left(H + \frac{u^2}{2g}\right)} = mS\sqrt{2gH + u^2}.$$

Let u be equal to 4 ft. per second, then taking $2g$ as approximately equal to 64, we have—

$$\frac{u^2}{2g} = \frac{16}{64} = \frac{1}{4} = 0.25 \text{ ft. to be added to H.}$$

D 2

31. *Flow of Water with Cylindrical Adjutages.*—The addition of a tube to an orifice in a thin plate gives a discharge larger than that through an orifice in a thin plate; but in order that it should produce this effect it is necessary that the water entirely fill the area of the external mouth of the tube, and this is generally the case when the length of the tube is two or three times greater than its diameter : if it be less than this, the fluid vein, which is contracted at the entrance, does not always enlarge so as to fill the interior of the tube ; the flow in that case takes place as if in a thin plate, and this is always the case when the length of the tube is less than the length of the contracted vein, which, as we have seen, is but half the diameter of the orifice, or even less.

32. The woodcut attached, which is a vertical section through the centre of the orifice and axis of the adjutage, serves to illustrate the action which takes place ; the fluid threads arrive at the orifice converging, and therefore the fluid will contract at the entrance. Experiments prove that this contraction is identical with that of the thin plate ; its position will, however, be internal with respect to the mouth of the tube attached. Beyond the section of

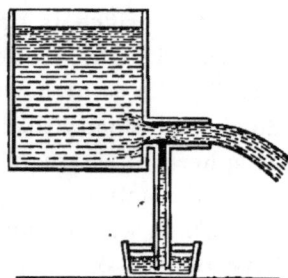
Fig. 12.

contraction, however, the attraction of the sides of the tube occasion a dilatation of the fluid vein ; the threads follow these sides, and issue parallel to each other and to the axis of the tube. The part which is darkened in the woodcut shows the space in which a partial vacuum is formed around the *vená contractá:* that such is the case is proved by this simple experiment ; a glass tube is inserted, air tight, in the side of the adjutage, the other end being placed under the free surface of the water contained in a vessel at a lower level; when the discharge takes place the water is seen to rise in the glass tube, to about ⅔ths of the charge, affording a measure of the degree of vacuum formed in the adjutage.

33. TABLE *showing the Increase in the " Coefficient of Contraction" by the Cylindrical Adjutage.*

Observers.	Adjutage.		Charge.	Coefficient.
	Diameter.	Length.		
	Feet.	Feet.	Feet.	
Castel, .	0.0508	0.1312	. 0.656	0.827
Do.	0.0508	0.1312	1.574	0.829
Do.	0.0508	0.1312	3.247	0.829
Do.	0.0508	0.1312	6.560	0.829
Do.	0.0508	0.1312	9.938	0.830
Bossut, .	0.0754	0.1771	2.132	0.788
Do.	0.0754	0.1771	4.067	0.787
Eytelwein,	0.0852	0.2558	2.361	0.821
Bossut, .	0.0885	0.1344	12.628	0.804
Do.	0.0885	0.1771	12.693	0.804
Do.	0.0885	0.3542	12.857	0.804
Venturi, .	0.1344	0.4198	2.886	0.822
Michelotti,	0.2656	0.7084	7.150	0.815
	Square.			
Do.	. 0.2656	0.7084	12.464	0.803
Do.	0.2656	0.7084	22.008	0.803
			Mean Coefficient, 0.817	

34. The mean of these coefficients gives 0.817, its value is generally taken as 0.82, so that we have the following formulæ:—

$$Q = 0.82 \, S\sqrt{2gH} = 6.56 \, S \sqrt{H};$$

and if d be the diameter of a circular orifice—

$$Q = 0.82 \, \frac{\pi}{4} \, d^2 \sqrt{2g} \sqrt{H} = 5.152 d^2 \sqrt{H}.$$

35. In the case when the jet issues with the tube full, in threads parallel to the axis of the orifice, and when, consequently, the section is equal to that of the orifice, the diminution of the discharge can only occur from a diminution of the velocity; and the ratio of the actual to the theoretic discharge is the same as that of the actual to the theoretic velocity.

TABLE *showing the Identity of the "Coefficients of Discharge" and Velocity, with the Cylindrical Adjutage.*

Observers.	Coefficient of the	
	Velocity.	Discharge.
Venturi,	0.824	0.822
Castel, .	0.832	0.827
Castel, .	0.832	0.829
Mean,	0.829	0.826

The three quantities measured were the "charge" on the centre of the tube, the velocity computed by measuring ordinate and abscissa, as in § 8, and the volume discharged. The velocity due to the charge, compared with that so computed, gives the second column, and the product of the area of the tube into the velocity due to the charge, compared with the discharge, gives the third; that is—

$$V (= \sqrt{2gH}) : \text{computed Velocity}, :: 1 : 0.824, \text{ and}$$
$$S \times V : \text{Discharge} \qquad :: 1 : 0.822.$$

We must, therefore, conclude that the velocity of a jet of water at the extremity of a cylindrical adjutage is equal to 0.82 of that due to the charge, and that the head due to that velocity is but 0.67 of the actual head of the reservoir; that is $(0.82)^2$, because the heads or charges are as the squares of the velocities.

36. As to the cause of this increase of the coefficient from 0.62 to 0.82, D'Aubuisson ascribes it to the attraction of the sides of the tube and the divergence of the fluid threads : after they have come in contact with the sides they are forcibly retained by some such attraction as that which causes the rise of fluids in capillary tubes : by this same force the outer threads draw after them the inner, and so all the vein issues with a full tube, and passes with an increased velocity through the contracted section. The immediate cause is the contact ; and every circumstance which favours that tends to produce an augmentation of the coefficient.

37. *Flow] of Water through Conical Converging Adjutages.*—Conical adjutages, properly so called, that is, those which are slightly converging to a point exterior to the reservoir, augment the discharge still more than the preceding. They give jets of great regularity, and throw the water to a greater distance or height, and are hence frequently used in practice : the effects vary with the angle of convergence of the sides.

Two distinct contractions of the fluid vein take place with this adjutage—one internally, or at the entrance of the adjutage, which diminishes the velocity due to the charge ; the other at the exterior ; in consequence of which the true section of the fluid vein is slightly less than the area of the external mouth of the adjutage.

If, therefore, we put S for the section of the external orifice, V for the velocity due to the charge, the actual discharge will be expressed by $nS \times n'V = nn'SV$, the two coefficients n and n' must be found by experiment, n being the ratio of the section of the fluid at its least diameter to that of the orifice, or the *coefficient of the exterior contraction*, and n' that of the actual velocity to the theoretic, or the *coefficient* of *the velocity*, and nn', their product, is the ratio of the actual discharge to the

theoretic, or the *coefficient* of *the discharge.* The knowledge
of these two last is of practical importance in the case
of jets of water, as in fountains and fire-engines.

38. In order to determine the coefficients above
mentioned, and especially to ascertain the angle of convergence that gives the maximum discharge, experiments were undertaken with a number of adjutages

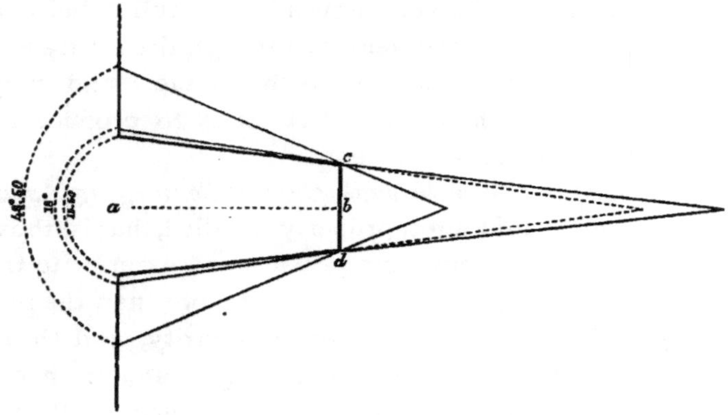

Fig. 14.

successively, in all of which the diameter of the orifice
of final issue *cd*, in the wood engraving, and the length
of the adjutage *ab*, remained constant; but in each experiment the diameter of entrance, and consequently the
angle of convergence, were altered. The flow of the
water was produced under different charges with each of
these varied adjutages.

At every experiment the discharge was determined
by actual gauging, and the velocity of issue by the
method of the parabola given above (§ 8). The discharge,
divided by SV, gave the product nn' and the observed
velocity divided by $V (= \sqrt{2gH})$, gave n'.

The series of the numbers nn' showed the discharge
corresponding to each angle of convergence, and consequently the angle of maximum discharge, and the

series of n', marked the progression by which the velocities increased.

39. The same adjutage, under charges which varied from 0.69 feet to 9.94 feet, or from 1 to 14, always gave discharges proportional to $\sqrt{H}$, and therefore the coefficient, or nn', has been, q, p, the same also. A very small increase may be observed with the higher charges. With respect to the coefficients of the velocity, they also should have been found constant but for the resistance of the air. Now, this resistance diminishing the throw of the jet, and that in proportion as the charge is greater, we should expect in the coefficients calculated from it a decrease augmenting with the charge—although, at the same time, there was no actual diminution in the velocity with which the fluid issued, or tended to issue.

TABLE *showing the " Coefficients with Conical Converging Adjutages," the Angle of Convergence being that giving the maximum Discharge, as determined in the next Table.*

Charge.	Coefficient of the	
	Discharge $= nn'$.	Velocity $= n'$.
Adjutage, . . . 0.0508 feet diameter.		
Feet.		
0.705	0.946	0.963
1.584	0.946	0.966
3.253	0.946	0.963
4.893	0.947	0.966
6.579	0.946	0.956
9.938	0.947	
Adjutage, . . . 0.0656 feet diameter.		
0.692	0.956	0.966
1.584	0.957	0.968
3.263	0.955	0.965
4.913	0.956	0.962
6.586	0.956	0.959
9.938	0.957	

Let us, in the next place, compare together the coefficients both of the discharges and of the velocities obtained, with the different adjutages of one and the same series,—adjutages which only differed in the angle of their convergence. Each coefficient is derived from a mean of five or six experiments taken with different charges, very nearly the same as those put down in the preceding Table.

TABLE *showing the Variation of the Coefficients of Discharge and Velocity with Conical Converging Adjutages at different Angles.*

Angle of Con-vergence.	Coefficient of the		Angle of Con-vergence.	Coefficient of the	
	Discharge.	Velocity.		Discharge.	Velocity.
Diameter, . . 0.0508 feet.			Diameter, . . 0.0656 feet.		
0° 0'	0.829	0.830			
1° 36'	0.866	0.866			
3° 10'	0.895	0.894	2° 50'	0.914	0.906
4° 10'	0.912	0.910			
5° 26	0.924	0.920	5° 26'.	0.930	0.928
7° 52'	0.929	0.931	6° 54'	0.938	0.938
8° 58'	0.934	0.942			
10° 20'	0.938	0.950	10° 30'	0.945	0.953
12° 4'	0.942	0.955	12° 10'	0.949	0.957
13° 24'	0.946	0.962	13° 40'	0.956	0.964
14° 28'	0.941	0.966	15° 2'	0.949	0.967
16° 36'	0.938	0 971			
19° 28'	0.924	0.970	18° 10'	0.939	0.970
21° 0'	0.918	0.971			
23° 0'	0.913	0.974	23° 4'	0.930	0.973
29° 58'	0.896	0.975	33° 52'	0.920	0.979
40° 20'	0.869	0.980			
48° 50'	0.847	0.984			

The facts established in the first Table, Diameter, 0.0508 feet, are represented graphically in the two following engravings; the upper referring to the discharges, the lower to the velocities. In each the horizontal line indicates degrees, extending from 0° to 48°, and having 0.3 inch equal to 4°. On this line the degrees of convergence of the adjutages are laid off as abscissæ from 0°, the coefficients corresponding being laid off on the dotted vertical lines as ordinates; the scales for these, in which 1.00 = 4 inches nearly, are given at the left-hand side commencing with 0.80, which is set to the horizontal line, as none of the coefficients were less, so that the datum line in each engraving is 3.2 inches below that given through 0.80. The curve line A, B, C, is drawn through the termination of the ordinates of discharge, and A′, B′, C′, through those of the velocity.

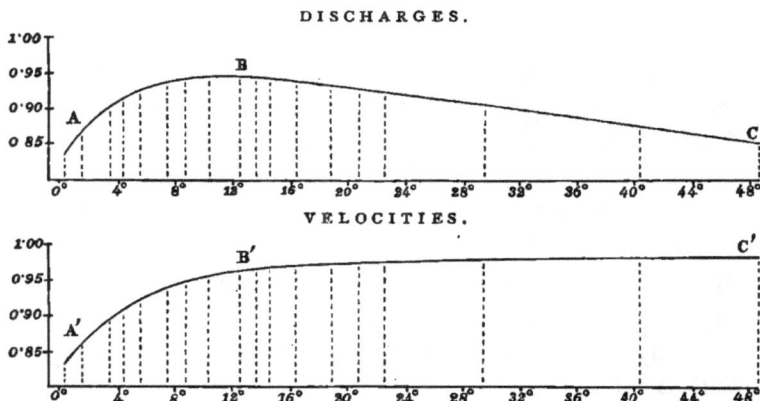

Fig. 14.

40. It follows, both from the Table and the wood engravings, 1st, That, for the same orifice, and with the same constant charge, the actual discharge, commencing with 0.83 of the theoretical, increases gradually, in proportion as the angle of convergence increases, up to $13\frac{1}{2}°$, near B, at which the coefficient of discharge is 0.95: beyond this angle it diminishes at first slowly, as do all variables about the maximum. At 20° the coefficient is

yet as high as 0.92 : subsequently the diminution becomes more and more rapid, and terminates as low as 0.65, which is the coefficient of discharge through a thin plate, —this last being the ultimate position of converging adjutages, that, namely, in which the angle of convergence has attained its maximum, or 180°. Thus, then, we have for the maximum discharge an angle of convergence of between 13° and 14°.

2ndly. In looking at the coefficients and ordinates of velocity, we see them also increasing from 0°, nearly as those of the discharge, up to 10°; after that they increase more rapidly; and when beyond the angle of maximum discharge, while the coefficients of discharge diminish, they continue to augment and approach their limit of unity. They are very nearly equal to unity at 50°, and even at 40°, are not far from it. Thus, conical adjutages may, by varying the angle of convergence, be made to form a series or progression, whose first term is the cylindrical adjutage, and last the orifice in a thin plate; the velocity of projection, increasing with the angle of convergence, will vary from that of the tube additional, up to that of the simple orifice in a thin plate; that is to say, from $0.82 \times \sqrt{2gH}$ up to $1 \times \sqrt{2gH}$.

3rdly. If we compare the coefficients of discharge with those of the velocity,—that is, the successive values of $n \times n'$ and n', and divide the former by the latter,— we shall have the values of n, or the coefficients of the exterior contraction. From the angle 0° up to 10°, n is sensibly equal to 1, and, consequently, no such contraction was present in the experiment; and, notwithstanding the convergence of the sides, the fluid particles issued, q, p, parallel to the axis of the cone. Beyond 10°, however, the contraction becomes apparent; it reduces more and more the section of the vein, and terminates by bringing it to an equality with that of the orifice in the thin plate, as is shown here :—

TABLE *showing the Value of n, the Coefficient of Exterior Contraction, with different Angles of Convergence.*

Angle.	n, or $\frac{nn'}{n'}$.
8°	1.00
15°	0.98
20°	0.95
30°	0.92
40°	0.89
50°	0.85
180°	0.65

In these experiments the length of the conical adjutage was fixed at about 2½ times the exterior diameter, as shown in Fig. 14; thus it was 0.1312 feet for those of 0.0508 feet, and 0.1640 feet for those of 0.0656 feet, in order to avoid, as far as possible, complicating the results with the effect of the friction against the sides, in this following the analogy of the cylindrical adjutage,

Fig. 15.

in which experience proves that, with respect to discharge, they produce their full effects most certainly when the length equals 2½ times the diameter.

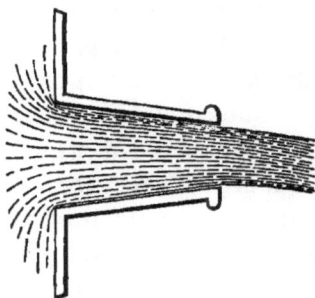

41. As to those very large conical adjutages, or rather truncated pyramidal tubes, which in some manufactories on the Continent discharge water upon mill-wheels, three very valuable experiments, made at a mill on the canal of Languedoc, are given by the engineer,

Lespinasse. They were, in this case, formed by the sides of a rectangular pyramid, whose length was 9.59 feet; rectangle of large end, 2.4 feet by 3.2 feet; at the smaller, 0.443 feet by 0.623 feet.

The opposite faces made angles of 11° 38′ and 15° 18′; the charge was 9.6 feet :—

TABLE *showing the " Coefficient" of Discharge with very large Converging Mill-sluices.*

Discharge.	Coefficient.
Cubic Feet.	
6.767	0.987
6.692	0.976
6.714	0.979

We see, then, how very little such adjutages diminish the discharge : that which they give is only one or two hundredths below the theoretic discharge.

42. *Conical Adjutages Diverging.*—This adjutage, of all others, gives the largest discharge. It may be described as a truncated cone attached to a reservoir by its smaller diameter, and of which the exterior mouth is consequently greater than that of the entry of the water. Although not much in practical use, they present phenomena of such interest as to deserve some notice.

The property they have of increasing the discharge was known to the ancient Romans : some of the citizens, to whom had been granted the privilege of having a certain quantity of water from the public reservoirs, found, by using these adjutages, the means of increasing their supply; and the fraud became so extensive that their use was forbidden by law, except when the distance from the reservoir was not less than about 52 feet. Venturi is the experimenter to whom we are chiefly indebted for information respecting this particular adjutage.

43. Those which he made use of carried a mouth-piece, ABCD, not unlike the form of the contracted vein, AB being equal to 0.1332 feet, and CD equal to 0.1109 feet. The body of the adjutage varied in length and in its divergence : this last was measured by the angle contained between the sides EC and FD, supposed pro-

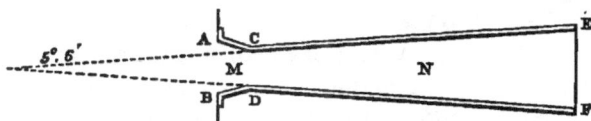

Fig. 16.

longed until they meet. These adjutages were attached to a reservoir maintained at a uniform level; the flow took place under a constant charge of 2.89 feet; and the time required to fill a vessel of 4.838 cubic feet was observed.

The following Table gives the result of his principal observations, premising that the time corresponding to (unity as a coefficient, that is to) the theoretic velocity, was 25.49 seconds :—

TABLE *showing the Variation of the "Coefficient" of the Discharge with Conical Diverging Adjutages at different Angles and Lengths.*

Adjutage.		Time of Flowing.	Coefficient.
Angle.	Length.		
	Feet.	Seconds.	
3° 30'	0.364	27.5	0.93
		25.49	1.00
4° 38'	1.095	21	1.21
4° 38'	1.508	21	1.21
4° 38'	1.508	19	1.34
5° 44'	0.577	25	1.02
5° 44'	0.193	31	0.82
10° 16'	0.865	28	0.91
10° 16'	0.147	28	0.91
14° 14'	0.147	42	0.61

Venturi has drawn the conclusion that the adjutages of maximum discharge should have a length of nine times the diameter of the smaller base, and an angle of divergence equal to 5° 6′: it is represented in the wood-cut, Fig. 16. This, he adds, would give a discharge 2.4 times greater than the orifice in a thin plate, and 1.46 times greater than the theoretic discharge.

44. *Flow of Water under very small Charges.*—When the charge over the centre of the orifice is very small compared with the vertical depth of the orifice itself, the mean velocity of the different threads of the fluid vein—that is to say, the velocity which, being multiplied by the area of the orifice, gives the actual discharge—is no longer that of the central thread. It differs from it in proportion as the charge is less: its true value will be about the hundredth part less when the charge is equal to the depth of the orifice, and by about the thousandth part when equal to three times that depth. Let us examine what theory teaches on this point; and first, of the law which it assigns for the velocity of the fluid threads in proportion as their depth below the surface of the water increases.

The italic capital *H* is used for the depth from the surface to the sill or bottom line; the italic *h* for the depth to the top of any rectangular orifice; and the capital H for the mean depth, or $\dfrac{H + h}{2}$.

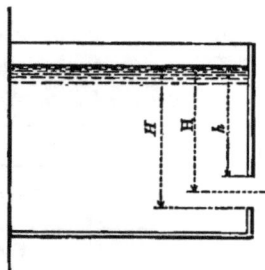

Fig. 17.

45. *Velocity of any Fluid Thread whatever.*—Let AA, Fig. 18, be the level of the surface of water in a vessel, and upon the face AB—which, for greater simplicity, we suppose vertical—let us imagine a series of very small orifices placed one below the other, and of which that at B is

the lowest, and putting H for the height AB, the velocity of the jet issuing from B will be expressed by $\sqrt{2gH}$; making BC equal to this quantity, it will represent this velocity; for any other point P, taken at a depth equal to AP or x, the velocity of issue will be represented by the line PM $= \sqrt{2gx}$, and calling this y, we shall have $y = \sqrt{2gx}$, or $y^2 = 2gx$.

Through all the points M so found drawing a curve line, it will, from the above equation, be a parabola having $2g$ or 64.4 feet for its parameter; and thus we have this proposition :—The velocity of a fluid thread issuing at any depth is equal to the ordinate of a parabola whose parameter is equal to $2g$, and the depth the abscissa.

46. Let us next suppose that this series of orifices over each other was continuous, forming a rectangular slit, whose width was l, and seek now the Discharge in

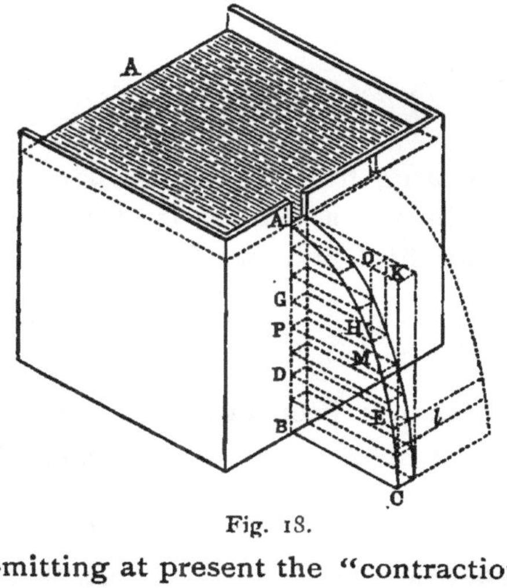

Fig. 18.

this case, omitting at present the "contraction." Suppose this opening divided into elementary rectangles by horizontal lines, the volume of water which will issue from each of them in one second will be equal to the

E

volume of a prism whose base is the elementary rectangle, and height the velocity, or ordinate, corresponding. The sum of all the prisms will also be equal to a single prism whose base is the parabolic segment ABCMA, and height l, the width of the opening.

From a property of the parabola, this segment is ⅔rds of the rectangle ABCK, whose area AB × BC, as shown above, is equal to $H \times \sqrt{2gH}$; thus the discharge for the rectangular opening, whose height is H and width l, is

$$\frac{2}{3} \times l \times H \sqrt{2gH} = \frac{2}{3} l \times \sqrt{2g} \times H \, . \, \sqrt{H}.$$

47. Let us now seek to determine the discharge through a rectangular orifice opened in the same face, but only from B to D, and with the same width l. Let h = AD, then the discharge of the opening from A to D will be—

$$Q = \frac{2}{3} l \times h \times \sqrt{2gh} = \frac{2}{3} l \sqrt{2g} \times h \sqrt{h}.$$

Now the discharge from the rectangular orifice, BD × l, will be the difference of those from the openings AB and AD, each into l, and therefore

$$Q = \frac{2}{3} \times l \sqrt{2g} \times \left(H \sqrt{H} - h \sqrt{h} \right).$$

That which has been established in §. 13—namely, $Q = S \sqrt{2gH}$, is, substituting for S its value, $S = l (H - h)$,

$$Q = l \sqrt{2g} \, . \, \sqrt{H} \, . \, (H - h),$$

on the supposition, very nearly correct, that the velocity at the mean depth was the mean velocity.

48. *Mean Velocity.*—Let us now determine the mean

velocity (§ 44), and first, that of the rectangular opening up to the surface. Let G be the point from which
issues the thread with this mean velocity. If we make
AG = z, it will be expressed by $\sqrt{2gz}$, and this being
multiplied into the area of the opening, $l \times H$, must give
the total discharge, which we have already found equal

to $\frac{2}{3}.l.H\sqrt{H}\sqrt{2g}$, we shall therefore have—

$$l \times H\sqrt{2g} \times \sqrt{z} = \frac{2}{3} l \times H \times \sqrt{2g} \times \sqrt{H},$$

dividing each side by the common factors $l \times H \times \sqrt{2g}$,
we have, squaring both sides,

$$z = \frac{4}{9} H;$$

and, therefore, $V = \sqrt{2g\frac{4}{9}H} = \frac{2}{3}\sqrt{2gH}.$

Thus it appears that the mean velocity is equal to two-
thirds of the velocity of the lowermost fillet; and so GH,
which represents the former, is $\frac{2}{3}$rds of BC, which, in like
manner, represents the latter.

For the case of the rectangular orifice, whose depth
is BD or $H - h$, we shall, in like manner, have the area
expressed by $(H - h) \times l$, and the discharge, making z'
the height due to the mean velocity, by—

$$(H-h) \times l \times \sqrt{2gz'} = \frac{2}{3} l \times \sqrt{2g} (H\sqrt{H} - h\sqrt{h}).$$

Dividing both sides by $l \times \sqrt{2g}$, we have—

$$(H-h) \times \sqrt{z'} = \frac{2}{3}(H\sqrt{H} - h\sqrt{h}),$$

dividing by $H - h$, and squaring—

$$z' = \frac{4}{9}\left(\frac{H\sqrt{H} - h\sqrt{h}}{H - h}\right)^2.$$

and hence, the mean velocity,

$$V = \sqrt{2g} \times \frac{2}{3} \times \frac{H\sqrt{H} - h\sqrt{.h}}{H - h}.$$

In order to have a clear idea of the difference in the discharges computed by the two formulæ, that, namely,
which multiplies the velocity at the mean depth into the
area of the orifice, which is the one almost always employed, and the true formula given in § 47, the Table,
page 54, was calculated, in which the area of the sluice,
as opened, is supposed to be 2ft. deep by 4 ft. wide, and
placed in ten different positions as to depth from the surface. In the first, the upper line of the orifice is on the
level of the water surface, and therefore $h = 0$; in the
second, 2 ft. below it, and so on successively 2 feet deeper,
until $h = 18$, and $H = 20$; the several values of H will
therefore be 1, 3, 5, &c., up to 19.

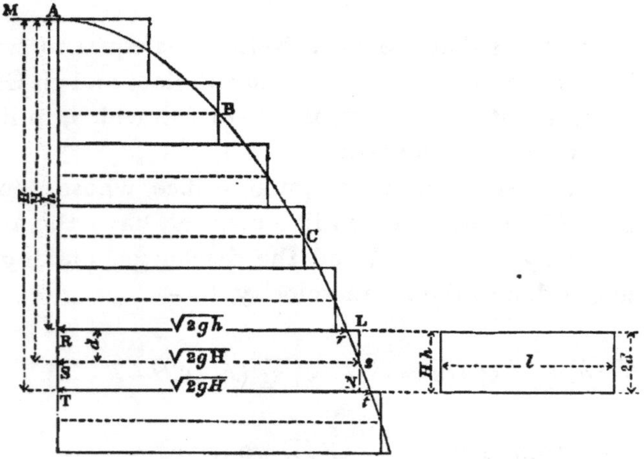

Fig. 19.

The wood engraving illustrates this arrangement as

far as the seventh position, and also shows graphically
the reason of the difference of the results in calculating
by the two formulæ. The level of the water surface is
supposed to be on the line MA, and the vertical line
ARST, the edge of the wall or plate seen in the transverse section, in which the orifice or sluice-way is supposed to be opened; RT = 2 ft. is the sixth position in
which the opening is placed, and, therefore, $h = 10$, $H = 12$,
and $H = 11$. The curve line is the parabola drawn through
all the points r, s, t, found by plotting the lines $Rr = \sqrt{2gh}$,
$Ss = \sqrt{2gH}$, &c., which, therefore (as in Figure 18, page
49), represent the spaces that would be traversed, in the
horizontal, in one second by a particle of water issuing
at that depth.

Hence, the area of the parabolic zone, $RrtTR$ multiplied into l, the width of the sluice (which is shown in
elevation at the right-hand side), will give the true volume
of water discharged in one second at that depth.

The simple formula, § 13, gives the volume as the rectangle RLNT multiplied into the same width l. The
vertical LN being drawn through s, for Ss represents the
velocity at the mean depth $H = (H + h) \div 2$.

If through the point s we draw a tangent to the curve,
or any right line terminated by the parallels RL and TN,
produced if necessary, the trapezium so formed is equal
to the rectangle RLNT, and the more nearly the portion
of the curve rst approaches a right line, the more nearly
identical are the results of calculating the discharge by
the two formulæ. Now, as the curve rst is always concave towards the vertical AT, the true formula gives
a less result than that in § 13. It may also be observed,
that as we descend, the portion of the curved line between the equidistant parallels is more nearly rectilinear,
as at B, C, and s successively, and therefore the results

also gradually approach equality, and Ss is more nearly the arithmetic mean between Rr and Tt.

TABLE *showing the Error of calculating the Discharge by the usual Approximate Formula,* § 13.

The Letters h, H, *H, are used as in* § 44.

1			2	3	4	5	6
					Differ-	Col. 4	Col. 4 as
			$\sqrt{2g}.\sqrt{H}.(H-h)$	$\sqrt{2g}.\tfrac{2}{3}(H\sqrt{H}$	ence of	$\times 0.62$	per cent.
h	H	H		$-h\sqrt{h})$	2 and 3	$\times 4$.	of Col. 2.
0	1	2	16.000	15.085	0.915	2.270	5.7
2	3	4	27.712	27.581	0.131	0.325	0.475
4	5	6	35.777	35.717	0.060	0.149	0.17
6	7	8	42.332	42.295	0.037	0.092	0.087
8	9	10	48.000	47.975	0.025	0.062	0.052
10	11	12	53.066	53.047	0.019	0.047	0.036
12	13	14	57.689	57.674	0.015	0.037	0.026
14	15	16	61.968	61.956	0.012	0.030	0.019
16	17	18	65.970	65.960	0.010	0.025	0.015
18	19	20	69.742	69.734	0.008	0.020	0.0116

The first column gives the successive depths below the surface; the second, the corresponding area of the rectangle Ss × RT; the third, that of the parabolic zone between the same parallels; the fourth, the difference between these two *areas ;* and the fifth, the difference of the *volumes,* given by each formula in one second through the opening 2 ft. by 4 ft., the coefficient of contraction being supposed 0.62; in the sixth the percentage of error by using the simpler formula is shown.

Let us, for further illustration, suppose that a reservoir has on one side a rectangular orifice, 2 ft. wide and 1.4 ft. deep; the surface of the water being maintained at a constant height above the sill of 2 ft. = H; required the height z' due to the mean velocity of the flow. As $H = 2$ ft., we have $h = 2 - 1.4 = 0.6$ ft., and

$$(a) \quad . \quad . \quad . \quad z' = \frac{4}{9}\left(\frac{2\sqrt{2} - 0.6\sqrt{0.6}}{2 - 0.6}\right)^2 = 1.2664 \text{ ft.} ;$$

and as $(H + h) \div 2 = 1.3$ ft., the difference 0.0336 feet shows the smallness of the error of the simpler formula when $H = 2$ feet; had it been 6 feet, all else being the same, then from (a), $z' = 5.29$ feet, and as $(H + h) \div 2 = 5.3$ feet, the error is now only 0.01 feet. The respective mean velocities are in the first case $\sqrt{2g \times 1.2664} = 9.0308$ feet and $\sqrt{2g \times 1.3} = 9.15$ feet; in the latter they are 18.4556 feet and 18.475 feet.

49. *The Charge must be measured from the Level of Still Water.*—It may be proper here to observe, that during the flow of water through an orifice the surface of the water in the reservoir takes a curved form for a certain distance, and bends towards the side in which the orifice is opened; so that the vertical height of the surface above any point in the orifice, estimated beyond the place at which this curvature commences, is greater than that immediately over the orifice. The former of these heights or charges is that which must be introduced into the formula for the discharge, as will be mentioned in § 55. By overlooking this, as has been frequently done, an error is introduced into the calculation of the discharge, which is thus deficient; and in some cases, though very rarely, it may amount to a tenth part. The error diminishes as the charge increases, and, according to Poncelet, becomes insensible when the charges exceed 0.5 feet to 0.66 feet. This, however, refers to the small experimental orifices used by him. D'Aubuisson states that he measured the depression of the surface of the water at the lock-gates of the canal of

Languedoc, and found it, when both sluices were open, from 0.1312 feet to 0.1640 feet.

50. *Coefficient for the Reduction of theoretic to actual Discharges.*—The discharges which have been spoken of (§§ 46, 47) are the theoretic discharges. In order to reduce them to the actual discharges, it is necessary to multiply them by coefficients deduced from experiment. MM. Poncelet and Lesbros have determined their values, as in the Table following :—

TABLE *giving the " Coefficients" for the Reduction of the Discharge, through Orifices calculated by the true Formula, to the actual Discharge.*

Charge on the Centre.	HEIGHT OF THE ORIFICES.					
Feet.	Feet. 0.656	Feet. 0.328	Feet. 0.164	Feet. 0.098	Feet. 0.065	Feet. 0.032
0.032						0.712
0.065				0.644	0.667	0.700
0.098				0.644	0.663	0.693
0.131			0.624	0.643	0.661	
0.164			0.625	0.643	0.660	
0.196		0.611	0.627	0.642		
0.262		0.612	0.628	0.640		
0.328		0.613	0.630	0.638		
0.393	0.592	0.614	0.631			
0.492	0.597	0.615	0.631			
0.656	0.599	0.616	0.631			
0.984	0.601	0.617				
1.640	0.603	0.617				
3.281	0.605					

51. The numbers given above are the true coefficients of contraction of the fluid vein, or coefficients of reduction of the theoretic to the actual discharge; for theory gives no other general formula for the flow of water through orifices than—

$$Q = m \frac{2}{3} l \times (H \sqrt{H} - h \sqrt{h}) \times \sqrt{2g}.$$

That which has been established in § 13, namely, $S = \sqrt{2gH}$, in which $H = \frac{1}{2}(H + h)$ is only applicable in particular cases, but which, no doubt, are of very frequent occurrence in practice; in which H is three or four times greater than $H - h$. In all others it is to a certain extent erroneous; and the coefficients adapted to that formula, and which it has served to determine, are so also: those, namely, in the Table § 21, which are above the darker type in each column. The coefficients that are below these lines, though determined by the aid of the formula, $S\sqrt{2gH}$, are coincident with those derived from the general formula, and are correct. Moreover, in the equation, $Q = mS\sqrt{2gH}$, the error of the coefficient m is compensated by that of the formula itself, and discharges found by it are sensibly identical with those of the true general expression, and as it has the advantage in simplicity, it is generally employed in every case.

52. To illustrate the practical effect of the use of each formula, let it be required to find the discharge through a rectangular orifice $0^m.30$ wide and $0^m.15$ high, and having a charge of only $0^m.05$ upon the upper edge: we have $H = 0^m.05 + 0^m.15 = 0^m.20$ and $l = 0^m.30$. The charge upon the centre is therefore $0^m.125$, and the corresponding coefficient, according to the Table, § 21, is nearly 0.603, —the mean between 0.592 and 0.614. Thus the discharge will be—

$$\frac{2}{3} \times 0.603 \times 0.30 \times 4.43\,(0.20\sqrt{0.20} - 0.05\sqrt{0.05}) = 0^{mnm}.0418.$$

The usual formula $Q = mS\sqrt{2gH}$, with $m = 0.592$, derived from the Table § 21, gives—

$$0.592 \times 0.30 \times 0.15 \times 4.43\sqrt{0.125} = 0^{mnm}.0417.$$

53. *Flow of Water over Waste-boards, Weirs, &c.*—If a rectangular opening with horizontal base be formed at the upper part of one of the sides of a basin, the water—supposed to be maintained at a constant level—will flow out in the form of a sheet over this base or sill : such are the waste-boards of canals and reservoirs, and also weirs in rivers, which extend across their course, so that the water must, when it meets with them, rise, and flow over the crest or summit.

The surface of the water, before it arrives at a waste-

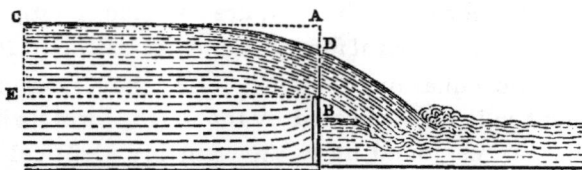

Fig. 20.

board, and from a point C, a short distance above it, is inclined in a curved line, CD, so that the depth, immediately above the sill, is not equal to AB, but only BD.

54. If we followed the usual theory, we should at once grant that the particles which followed the curve CD had the same velocity when arrived at D as if they had fallen freely from the height AD, and that all the particles in the vertical line under it would in like manner flow out with velocities due to their several heights, as referred to the point A : we should thus find, that in respect to the issuing velocity of the fluid threads, and also in respect to their number—which depends on the height BD,—and consequently in respect, also, of the discharge, that the case would be identical with that of a rectangular orifice closed on the upper side as far as D, and in which the water level extended without any curvature up to A; and, therefore, if we should put Q for the volume of water discharged in one second, l for the width of the opening,

and H and h for the charges,—the one on the lower, the other on the upper edge,—and lastly, m for the coefficient of the reduction of the theoretic to the actual discharge, we should have the expression, as in §47,—

$$Q = m \cdot \frac{2}{3} \sqrt{2g} \cdot l\, (H \sqrt{H} - h \sqrt{h}).$$

55. However natural it appears to proceed thus, facts show that the discharges are more exactly given by the supposition that the flow takes place as if from the whole height AB, the level of the water being supposed continued without curvature up to A. The case then would be identical with that of §46: for when $h = 0$ we should have—

$$Q = m \cdot \cdot \frac{2}{3} \sqrt{2g} \times l \times H \sqrt{H},$$

or since

$$\frac{2}{3} \sqrt{2g} = \frac{2}{3} \times 8.024 = 5.35,$$

we have

$$Q = m \times 5.35 \times l \times H \sqrt{H}.$$

The flow of water, then, over weirs or waste-boards is only a particular case of the flow by orifices in general; that, namely, in which the charge over the upper edge is equal to zero. MM. Bidone and Poncelet have shown this is so, and that the coefficients m, which suit ordinary orifices, are adapted to weirs also, when the flow is made under analogous circumstances.

We cannot, therefore, expect this coefficient to be constant; it varies, in fact, through a considerable range, from about 0.5 to 0.8, in circumstances, some of which will be recognised as nearly identical with those which influence the discharge through orifices. In determining it for weirs and overfalls it is now also necessary, in the first place, to be satisfied that the velocity of approach of the water in the channel supplying the overfall be not so great as to produce any appreciable increase in the discharge, and thus vitiate the result : whether the object be experimentally to determine the coefficient in the formula, or in practice to gauge a water-course : for either end it will appear further on that the transverse area of the sheet of water passing through the overfall should not exceed the fifth part of that of the channel.

And not only the *area*, but the *form* of the rectangle of the overfall, relatively to that of the transverse area of the channel, should observe certain limits, if we would expect to arrive at a true value for the coefficient. Thus,

Fig. 21.

let us suppose a rectangular opening cut in a thin plank or metal plate, the vertical centre lines of the channel and overfall coinciding, in plan and end view. The fluid threads, in all that part of the channel not immediately opposite to the opening, converge towards its edges ; the longitudinal sections, Figs. 20 and 21, in a plane perpendicular to the crest of the overfall, represent the contraction so far as the sill is concerned, the water below the line EB, in the engraving (page 58), bending upwards as it approaches the bottom line or sill B : the wood-engraving, § 62, which represents the plan of an overfall, shows the contraction that takes place at the sides, L being the width of the channel of supply, and *l* that of the overfall ; and this convergence, continuing after the fluid has passed the edge, produces, as in orifices, a contraction of the section of the sheet of water which diminishes the discharge.

Now if the form of the opening be altered, the area still continuing of the required proportion to that of the channel, i. e. ⅓th, so that in the two extreme cases, when, on the one hand, L = *l*, and therefore the depth of the sheet of water not more than ⅓th of that of the channel, we would have the *end* contraction suppressed; and on the other hand, if the depth of the overfall were equal to that of the channel, the length *l* being now reduced to ⅓th, we would have the *bottom* contraction suppressed, as with orifices in the woodcut, § 25. Moreover, in the intermediate rectangular forms of the overfall, in which both contractions come into operation, we may trace a difference of coefficients according to the different depths of the rectangles below the surface.

A floor or channel is sometimes attached either above or below the line of the sill, and vertically on either or both sides, above or below, alterations in the value of the coefficients resulting in each case. It must be evident also, that if instead of a thin plank or metal plate, the sill and sides were of considerable thickness, and not bevilled off, the sheet of water, especially if the depth be inconsiderable, and the velocity consequently small, would adhere to the surface, and the coefficient become modified, somewhat similar to the case when a floor is added below.

The effect of the contraction on the bottom line is evidently the same on each unit of the length; but the end contraction must be nearly constant with the same depth, however the length may be increased or diminished; and therefore we cannot anticipate that *m*, § 59, which includes the effect of both the end and bottom contractions, should be rigidly constant when the length of overfall varies: an attempt to express the effect of each in the formula is given in § 62.

56. *Effects of the Velocity of Approach.*—When establish-

ing the formulæ given above, § 13 and 47, it was assumed
that the water was free from any current above that point
at which the surface begins to be curved towards the sill;
but frequently the water arrives at this point with an
initial velocity; in this case we proceed as in § 30, for the
case of orifices properly so called : that is, we add to the
head due to the velocity (in the case when the water has no

appreciable velocity and which is now only $\frac{4}{9}$ *H, vide* § 48),

that which would generate the velocity with which it arrives at the edge.   Let u represent this velocity in ft. per
second, and h ft. the charge due to it, then—

$$\text{since } h = \frac{u^2}{2g} = \frac{1}{64.4} \times u^2,$$

we have $h = 0.01553\, u^2$, (as $\dfrac{1}{64.4} = 0.01553$),

representing the additional head in feet; and we shall
have, for the velocity, V, of the issuing water—

$$V = \sqrt{2g\left(\tfrac{4}{9}H + h\right)} = \sqrt{2g\left(\tfrac{4}{9}H + 0.01553\, u^2\right)},$$

$$= 8.024 \times \frac{2}{3}\sqrt{H + \frac{9}{4} \times 0.01553},$$

which may be reduced to—

$$5.35\,\sqrt{H + 0.035\, u^2}\,;$$

as $\dfrac{9}{4} \times 0.01553 = 0.035$; and consequently,—

$$Q = 5.35 \times m \times l \times H \times \sqrt{H + 0.035\, u^2}.$$

The quantity u represents the *mean* velocity of the
section of the water approaching the edge.   The exact
determination of it is, perhaps, impossible: but as its

value will be very little different from that at the *surface* (which can be readily determined by methods to be subsequently pointed out), we may, therefore, grant the truth of the equation, and modify the value of the coefficient which is to be determined experimentally. If, then, we put m' for this new coefficient, and w for the velocity of the surface of the water, we shall have—

$$Q = 5.35 \; m' \times l \times H \sqrt{H + 0.035 \; w^2}.$$

57. Reverting now to the formula, p. 59.

$$Q = m \times 5.35 \times l \times H \sqrt{H},$$

it remains to put it to the test of actual experiment. This expression for the discharge contains two quantities, which vary in the experiment,—the *width* of the overfall ; and a function of the velocity, that is, of the *charge*. Now, in order that the formula be well founded, it is necessary that the discharge be exactly proportional to each of these ; then, and then only, will the coefficients be constant. The degree in which they are so in the experiments will thus be a test of their being well founded in truth. The numerous experiments of M. Castel, engineer to the waterworks of Toulouse, are quoted by D'Aubuisson, and a description of his apparatus is also given.

At one extremity of a wooden canal, Fig. 23, page 65, 19.68 ft. long, and 2.427 ft. (= 2 ft. 5⅛ ins.) wide, and 1·804 ft. (= 1 ft. 9⅘ ins.) deep, he received the supply of water, and at the other he had the power of placing different thin plates of copper, in which the overfalls were cut out ; their width was increased gradually from 0.0328 feet (about ⅜th inch), up to 2.427 feet ; that is, the full width, the sill being maintained always 0.557 feet (about 6⅘th inch) above the bottom of the canal. The water discharged was received, for any required length of time, in a cistern lined with zinc, the capacity of which was 113 cubic feet : this was the gauge basin. The depths were marked with the greatest care on a vertical scale. The time that the water might take to reach any height was taken by a chronometer marking quarter seconds.

58. The charges or heights of the water above the sill of the overfall were successively increased from 0.0984 feet ($1\frac{3}{16}$ inch), up to 0.328 feet, and even to 0.787 feet (nearly $9\frac{1}{2}$ ins.) for the narrower overfalls. The most important, and at the same time the most difficult, point in the experiments, was to measure the charges exactly. In order to accomplish this, M. Castel placed over the centre of the canal, and parallel to its length, as in the wood-cut, a bar truly horizontal, which carried at every $0^m.05$

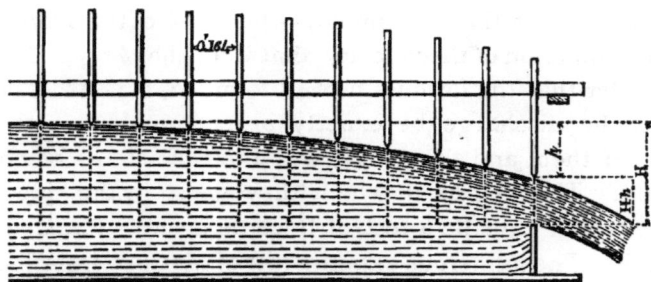

Fig. 22.

= 0.164 feet (nearly 2 inches), ten vertical brass rods, divided into millimetres, pointed and having a power of sliding up or down : on the edge of the guides was a vernier, which subdivided the rods into tenths of a millimetre.

When an experiment was made, the requisite quantity of water was admitted into the canal; and regulations for due constant supply being carefully attended to, he let down the rods, and put the points as accurately as possible in contact with the surface of the current.

Subtracting then their several readings from the vertical distance between the horizontal bar and the sill, he obtained the ordinates of the curve described by the particles of the fluid as they advanced to the centre of the overfall. The ordinates increased in proportion to their distance from it; and thus the greatest ordinate, or the true charge, H, was obtained; the smallest—that, namely, which was immediately above the sill—was $H - h$, or the thickness of the sheet of water at the moment of its passage over the edge of the waste-board.

59. After having to some extent exhausted all the observations which could be made with the canal of $0^m.74 = 2.427$ feet in width, M. Castel entered upon those in which he used a width only of $0^m.361 = 1.180$ feet, as in the woodcut, narrowing up the former by two partitions in plank, but whose length was only 7.347 feet,

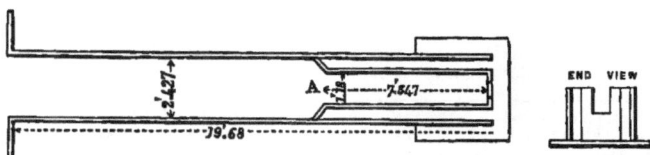

Fig. 23.

the total length of trough or channel being 19.68 feet. At the entrance A of this small canal, which was placed in the centre of the larger one, mentioned above, there was formed, with the large discharges, a minute fall, which would have introduced some slight modifications into the results obtained, if the partitions had been extended up to the extremity of the larger canal.

Upon both the one and the other M. Castel made a long series of experiments. Each observation was repeated once or twice: altogether he made 494. In every case, the values of Q, of l, and of H, were determined directly by measurement at each experiment; and from

F

them it was easy to deduce the value of the coefficient
m from the formula—

$$Q = 5.35\ m \times l\ H \sqrt{H}$$

for, dividing, we have—

$$m = \frac{Q}{5.35 \times l \times H \times \sqrt{H}}.$$

60. The mean values obtained from each charge and
each width of overfall are arranged in the two following
Tables.   No observations have been made in those cases
in which there are blanks.

TABLE *showing the "Coefficients" for Overfalls, in Castel's
Experiments, as depending on the Charge and on the Length.*

CANAL—2.427 feet wide (= 2 ft. 5⅜ inch).
COEFFICIENTS, the length of the Overfall being

Charge upon the Sill.	Feet. 2.42	Feet. 2.23	Feet. 1.96	Feet. 1.64	Feet. 1.31	Feet. 0.98	Feet. 0.65	Feet. 0.32	Feet. 0.16	Feet. 0.09	Feet. 0.06	Feet. 0.03
Feet. 0.78							0.595		0.615		0.639	
0.72							0.594		0.614		0.639	
0.65							0.596	0.594	0.614	0.629	0.640	0.670
0.59							0.595	0.594	0.613	0.628	0.641	0.672
0.52							0.595	0.592	0.613	0.628	0.642	0.674
0.45						0.603	0.593	0.592	0.612	0.628	0.643	0.675
0.39					0.621	0.604	0.592	**0.591**	0.612	0.628	0.645	0.678
0.32		0.657	0.644	0.631	0.621	0.604	0.593	**0.591**	0.612	0.627	0.648	0.687
*0.26	0.662	0.656	0.644	0.632	0.620	0.606	0.595	0.592	0.612	0.627	0.652	0.698
0.19	0.662	0.656	0.644	0.632	0.622	0.610	0.604	0.595	0.612	0.628	0.658	0.713
0.16	0.662	0.656	0.644	0.633	0.626	0.616	0.611	0.597	0.613	0.629	0.663	
0.13	0.662	0.656	0.645	0.636	0.632	0.623	0.619	0.604	0.614		0.669	
0.09	0.663	0.660	0.651	0.642	0.636	0.631	1.624	0.618				

Charge upon the Sill	CANAL—1.180 feet wide. COEFFICIENTS, the length of Overfall being									
	Feet. 1.180	Feet. 0.984	Feet. 0.656	Feet. 0.328	Feet. 0.301	Feet. 0.259	Feet. 0.164	Feet. 0.098	Feet. 0.065	Feet. 0.032
Feet.										
0.787				0.619			0.624	0.629	0.647	0.666
0.721				0.615	0.613	0.617	0.620	0.627	0.646	
0.656				0.611	0.608	0.614	0.618	0.626	0.645	**0.667**
0.590			0.633	0.608	0.606	0.610	0.616	0.626	**0.644**	
0.524			0.628	0.605	0.603	0.608	0.615	0.625	0.644	0.668
0.459		0.678	0.624	0.603	0.601	0.605	0.614	0.624	0.644	
0.393	0.700	0.666	0.620	0.600	0.599	0.603	**0.623**	0.646	0.674	
0.328	0.684	0.656	0.617	**0.598**	0.598	0.600	0.614	0.624	0.648	
*0.262	0.672	0.652	**0.616**	0.599	**0.597**	**0.599**	**0.613**	0.624	0.654	
0.196	0.669	**0.652**	0.617	0.600	0.597	0.600	0.613	0.626		
0.164	**0.667**	0.653	0.620	0.605	0.604		0.614			
0.131	0.668	0.653	0.624	0.613	0.611		0.613			
0.098	0.670	0.665	0.632	0.628	0.625					

61. Let us next examine the same simple and common formula, $Q = 5.35\ l \times H\sqrt{H}$; as to the proportionality of the discharges Q, to the function of the charge $H\sqrt{H}$. For this purpose let us take the twenty-two series of discharges obtained, each with the same width, but with different charges ; and reducing the discharges of each series to that which they would have been if one of them—that, for instance, obtained under a charge of 0.262 feet, marked with an asterisk in the Table—had been taken as unity.

TABLE *showing that the Discharges are proportional to*
$$H\sqrt{H} = H^{1.5}$$

Charges H.	Series of Discharges.			Series of	
	1.	2.	3.	$H\sqrt{H}$	$\dfrac{H\sqrt{H}}{-h\sqrt{h}}$
Feet.					
0.656		3.96	3.98	3.95	4.01
0.590		3.38	3.39	3.38	3.42
0.524		2.83	2.84	2.83	2.87
0.459		2.31	2.32	2.31	2.34
0.393		1.83	1.84	1.84	1.86
0.328	1.40	1.39	1.40	1.40	1.41
*0.262	1.00	1.00	1.00	1.00	1.00
0.196	0.650	0.652	0.650	0.650	0.643
0.164	0.494	0.498	0.495	0.494	0.486
0.131	0.354	0.381	0.354	0.354	0.345

In like manner reducing the series of values of $H\sqrt{H}$, and placing side by side these several series: we have the above Table. Under the head of "Discharges" are three columns; the two first having been given by the canal, 2.427 feet in width, with overfalls 1.968 feet, and 0.328 feet in length; the third being from the canal of half the above width, 1.180 feet, with an overfall 0.164 feet (nearly 2 inches) long.

It results from a comparison of the twenty-two series of discharges with one another, and with the series derived from the function $H\sqrt{H}$, that the numbers in the same horizontal line are very nearly identical; and this, notwithstanding so great a difference in the charges, and in the length of overfall, both in the canal 2.427 ft. wide, and that of one-half this width: thus the proportionality of the discharges to the function $H\sqrt{H}$ may be considered as established. In the highest charge, and especially in column 3 of Discharges, the velocity of approach seems to have increased the relative discharge: as is always observed when the area of the overfall is more than $\frac{1}{8}$th that of the channel: under column 2, with the smallest charge, the relative number is considerably in excess: irregularities of this kind caution us to avoid these small heads in practical gaugings.

Experiments to determine the index of H (which in the expression § 55 is taken at 1.5) are given by J. B. Francis, "Lowell Experiments," pp. 88, 95: Boston, U. S., 1855. The scale of these experiments was extremely large, and both the mechanism and method of observing evince the greatest care and accuracy. A large *constant* volume of water, not necessarily known in quantity, was discharged through a series of overfalls of different lengths, the extremes being (very nearly) 17 and 3.5 feet, the respective depths in one series being 0.3576 and 1.0447 feet. The average value of the index, resulting from seventy-one experiments, is 1.47. Although the volume of water discharged was not a necessary datum in calculating this index, it may be interesting to state, as showing the large scale at which this author operated, that the discharge due to the lengths and depths above mentioned is upwards of 12.5 cb. feet per second. Castel's largest experiments, § 60, with a length of 2.23 feet. and depth of 0.32, give about 1.4 cb. feet per second, only one-ninth of the above.

62. *Ratio of the Discharges to the Width of Overfalls.*—The coincidence of the formula with the experiments is not so close when it is the length of the overfall which we vary. The discharges, considered with reference to the variable *l*, do not follow the exact ratio of the widths, however natural it would appear to suppose beforehand that it would be so. Starting from an overfall the full width of the channel, they diminish with its decreasing width, but considerably faster than it up to a certain point, beyond which, on the other hand, they diminish less rapidly than the width. The columns of the Table appended serve to fix the ideas upon this point.

TABLE *showing the Ratio of the Discharges to the Width of Overfall.*

Canal 2.427 Feet wide.		Canal 1.180 Feet wide.	
Width.	Discharge.	Width.	Discharge.
1000	1000	1000	1000
919	911		
811	788	831	807
676	645		
540	507	554	507
405	371		
270	243	277	246
135	121	138	125
68	62		
40	40		
27	27		
13	14		

For the canal of 2.427 feet in width, we have twelve
widths of overfall, which are in the ratio of the numbers in the first column : in the second we have the order
in which the corresponding discharges decrease, and
which were obtained under charges varying from 0.196
feet to 0.328 feet. In the case of the canal 1.180 feet
wide, in which ten widths were experimented upon, those
only are stated which were nearly the counterparts of
the numbers in the other. The series of these ratios
point out the fact, that in the two canals the discharges
follow the same law in reference to the widths of the overfalls; but it is to the relative widths in their respective
channels, and not to the absolute widths.

The woodcut is intended to illustrate the effect of the
end contraction that takes place at overfalls and weirs ;
L is the width of the channel of supply, the water in
which is flowing in the direction of the arrow, and *l* the

PLAN.

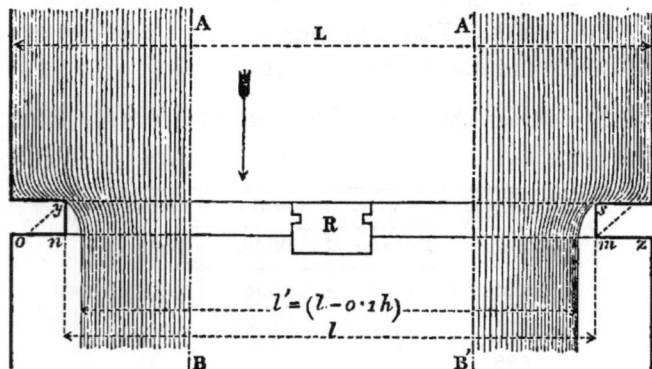

Fig. 24.

length of the overfall, at the place where the channel has
been contracted by the projecting edges *o, n, y,* and *z, m, s.*
The fluid threads are supposed to flow in lines parallel
to the sides, but as they approach the narrower opening

of the overfall, the length of which is *l*, those that move nearest the sides bend out of their course, converging towards its vertical edges throughout the whole depth; and this convergence, as in the *vená contractá* (§§ 15, 16), continues after the water has passed the points *y* and *s*, effectively shortening the length of the sheet of water passing over, which from *l* becomes *l'*, and thus diminishes the discharge. The convergence of the several lines is, however, less and less as their distance from the sides of the channel increases, so that at a short distance the direction of the fluid is quite unaltered; we may suppose the dotted lines AB and A'B' to include between them the portion of the channel and of the sheet of water uninfluenced by the convergence.

It has almost always been assumed in formulæ for the discharge of water over weirs, that the quantity is directly proportional to the length, the depth being the same. As this end contraction is a *constant* quantity, this supposition is not well founded when the length is considerable in proportion to the depth; a double length, as 2 *l*, giving more than a double discharge, and a length of ½ *l* giving less than half the discharge, because the same constant quantity is to be deducted from each to give the effective length when the head is invariable, and in like manner of other multiples of *l*. It is evident that the amount of this diminution of length from end contraction must increase with the depth; the exact law of its variation is not known, but experiments have pointed out that it is very nearly in direct proportion to the depth, which is of great importance in simplicity, when we attempt to introduce it into a formula for practical application. Thus, though neither this assumption, nor that of the index of *H* being constant, are perfectly correct—we find that they each lead to a result agreeing very closely with experiment.

The formula proposed by J. B. Francis (Lowell Experiments, p. 86), is

$$Q = C\,(l - bnh)\,h^a,$$

The quantity C, which this author calls the coefficient, stands for $\frac{2}{3}\sqrt{2g} \times m$ in the formula § 55. Other authors make the coefficient represent $\frac{2}{3}m$: as Morin, 2nd edit., in giving the Tables of Lesbros experiments. No error is caused in so doing; but a great want of uniformity results, and it requires care in comparing, and labour in reducing the Tables of different authors.

By the letter n is represented the number of end contractions, for we frequently find at weirs across streams (though not in experimental overfalls) that vertical posts, as at R, Fig. 24, p. 70, are placed along the crest, at nearly equal intervals, having vertical grooves, into which planks are dropped horizontally; these, being staunched against the crest of the masonry, raise the surface of the water, which in seasons of drought is necessary for many purposes; at each post the convergence and consequent contraction are identical with that at the two ends: at the least, then, n is equal to 2, that is, when the length is uninterrupted, and for every post it increases by 2: the constant multiplier or fraction of h is represented by b, and is such that $bh = \dfrac{l - l'}{2}$.

The coefficient b and index a are in one point distinct from m, inasmuch as they can be determined by experiments in which the actual discharge is not known in any particular case.

It is evident that this formula cannot be of general application, for if in the factor $l - bnh$, which represents the effective length, we should have $l = bnh$; then this effective length becomes 0, and the formula gives zero for the discharge, which is absurd; similarly, if $bnh > l$, the discharge would be negative. In weirs short in proportion to the depth flowing over, the effect of the end contraction cannot be considered independently of the length. It is found by experiment that, when the length equals or exceeds three times the depth, the formula applies; but in lengths less than this in proportion to the depth, it cannot be used with safety; the error increasing as the relative length of the weir diminishes. The end convergence influences the discharge to a certain distance from the end of the weir, as, suppose, from y to the line AB; or from s to A'B', if the whole length of the weir is greater than twice this distance, the effect of the end contraction is independent of the length; but if the length is less than twice this distance, the whole breadth of the stream is affected in its flow by the end contractions, and consequently the proposed formula would not apply. In practical construction it is nearly always possible so to proportion the weir, that the length may not be less than three times the depth upon it. In cases where the length of the weir is equal to the width of the channel, that is, $l = L$, there is no end contraction, so that we have $L = l = l'$, and the proportion of the length to the depth is not material.

The formula proposed by Francis, given above, deducts from the length l of the overfall a quantity presumed to represent the end contractions, and gives the value of l' the effective length. It is, giving b and a their numerical values,

$$Q = C\,(l - 0.05nh)\,h^{1.47}.$$

At the end of his second set of experiments Francis gives

$$Q = C\,(l - 0.1nh)\,h^{\frac{3}{2}}.$$

The use of the fractional index 1.47, not being so convenient, it has been increased 2 per cent., and the quantity deducted from l doubled, giving results agreeing most nearly with all his experiments. If $n = 2$, that is, if no vertical posts, as R, are fixed on the crest, then the quantities to be deducted from l are o.1h in the former, and o.2h in the latter formula.

63. *As to the Coefficients, first, for those having the same Widths.*—Resuming the consideration of Table, § 60, since the discharges are for the same widths sensibly proportional to $H\sqrt{H}$, when we omit extreme cases,—the coefficients ought to be very nearly constant, and we find they are so in the Table § 60. In strictness, when we take the coefficients of some one vertical column of the Table, we see them—commencing with the higher charges—decrease by very small degrees, in most of the experiments, down to a certain charge, beyond which they increase rapidly; thus we shall have at this particular charge, which is generally about o.328 feet, a minimum, given in the darker type.

And secondly, with the same constant charge, we may observe that the discharges decrease, at first more rapidly than the widths of the overfalls, and afterwards less so, it follows that under the same constant charge (the widths, commencing at a first width equal to that of the channel itself, being diminished) the coefficients de-decrease down to a certain point, beyond which they augment. Thus, here also there is a minimum; and it occurs when the width of the overfall is about the fourth part of that of the supplying channel; and so in both the horizontal and vertical lines of the Tables of coefficients, pages 66, 67, there is a minimum; we have, therefore, a general minimum. In its immediate neighbourhood, and for a certain distance, the variations are small; the coefficients in that part differ very little from one another, and may be regarded as constant. But in the other parts of the Table the differences rise to be considerable;

they exceed an eighth, or 12 per cent.; so that the discharge by overfalls cannot be given exactly with a constant numerical coefficient, m, in an expression of the form $H\sqrt{H}$. In practice, then, we would require the aid of very extended tables of coefficients, the preparation of which would demand many hundred experiments. However, the study of the direction in which the coefficients tend gives us the means of abridging this vast labour, and of determining a few simple rules suitable to the different cases which commonly present themselves.

64. *Coefficients and Formulæ to be employed.*—We have seen (§ 61) that the expression $l \times H\sqrt{H}$ should not be employed—on the one hand, when the charges were below 0.196 ft. (= 2⅜ ins.); and on the other, when the transverse area—that is, the depth multiplied by the width of the overfall—exceeds the fifth part of the area of the section of the water in the supplying channel, the initial velocity becoming then considerable. Between these limits, the expression above given can be employed with a coefficient,—variable, it is true, but only varying with the width of the overfall.

Commencing with a width equal to that of the canal itself, the coefficients diminish with the width of the overfall until it has reached the fourth part of that of the channel, and then they increase, although the widths are still decreasing; and what is very remarkable is, that the diminution of the coefficients follows the relative width of the overfall, compared with that of the canal, whilst the augmentation, which occurs afterwards, depends only on the absolute width. We have, consequently, four cases to distinguish relative to the coefficient to be used :—

First. In the neighbourhood of that common minimum we have mentioned, the variations of the coefficients are very trifling. According to the experiments made by Castel, from a width of overfall nearly equal to the third of that of the canal, which is supposed to exceed 0.984 feet, down to an absolute width of 0.164 feet, the coefficients do not vary more than from 0.59 to 0.61. Taking the mean, and remarking that—

$$5.35 \left(= \sqrt{2g} \times \frac{2}{3} \right) \times 0.60 = 3.21,$$

we shall have between the limits indicated above—

$$Q = 3.21 \times l \times H \sqrt{H}.$$

This formula furnishes the best mode of gauging small streams of water, as in the Examples appended.

Secondly. When the width of the overfall is at its maximum—that is, equal to that of the canal, extending from side to side, being then precisely similar to a weir, properly so called,—the coefficients present a remarkable steadiness under the different heads. M. Castel, in his experiments upon the canal of 2.427 feet, with an over-fall or barrier of a height equal to 0.557 feet, had no difference in the coefficients obtained with charges which varied from 0.098 feet to 0.262 feet, § 60. With an over-fall of 0.738 feet high, the coefficients have only ranged from 0.664 to 0.666 for charges of 0.101 feet up to 0.242 feet : at the mean he had 0.665. And since—

$$5.35 \times 0.665 = 3.55775,$$

we have, putting L for the width of the canal or length of the barrier—

$$Q = 3.558 \, L \times H \sqrt{H}.$$

This formula may be used with advantage in certain cases, even in large water-courses, and with charges of 0.131 feet and 0.098 feet; but to insure certainty in its use it is necessary that the charge be less than the third of the height of the barrier.

Thirdly. For widths of overfall comprised between that of the channel itself and the fourth part of the same, the coefficient of the expression $5.35\ l \times H \sqrt{H}$ will vary with the relative width,—that is to say, with the ratio of the width of the overfall to that of the canal of supply, and is given by the columns of this Table :—

TABLE *showing the Variation of the " Coefficient" with the relative width of the Overfall in the Formula* § 55.

Relative Widths.	Coefficients for Canal of	
	2.427 Feet.	1.180 Feet.
1.00	0.662	0.667
0.90	0.656	0.659
0.80	0.644	0.648
0.70	0.635	0.635
0.60	0.626	0.623
0.50	0.617	0.613
0.40	0.607	0.609
0.30	0.598	0.600
0.25	0.595	0.598

They have been formed by taking proportionals to the coefficients deduced from direct experiments, as given in § 60, a method which cannot here lead to any error. The coefficients determined by each canal have been given separately, in order to show that coefficients sensibly the same correspond to the same relative width, although the actual value of the widths was in one canal nearly double of that of the other, affording a proof that above 0.25, or the fourth of the width of the channel, the coefficients depend on the relative, and not on the absolute width of the overfall ; taking relative widths of 0.25, 0.6, and 0.8, and the mean coefficients, we have

$$Q = 5.35 \times \begin{Bmatrix} 0.596 \\ 0.624 \\ 0.645 \end{Bmatrix} l \times H \sqrt{H} = \begin{Bmatrix} 3.188 \\ 3.338 \\ 3.450 \end{Bmatrix} l \times H \sqrt{H}.$$

Fourthly. It is quite otherwise when this width falls below that of the fourth of the supplying canal : then, and when at the same time it is absolutely less than 0.262 feet or 0.196 feet, that of the canal has no further effect, and every particular width has its own coefficient. Thus in the canal of 1.180 feet, as well as in that of 2.427 feet,

the widths, . . 0.164 ft., 0.098 ft., 0.065 ft., 0.033 ft.,
have given, . . 0.61,   0.63,   0.65,   0.67,

as the respective coefficients in both canals.

65. *Observations on Formula of* § 54.—Having thus given in detail all that has reference to the simplest of the formulæ for the discharge of overfalls, let us consider the two others, and first, that of—

$$Q = 5.35 \, ml \, (H \sqrt{H} - h \sqrt{h}),$$

in which h represents the quantity AD, p. 58, Fig. 20, the surface of the fluid having become curved before its arrival at the overfall. A slight inspection of the last column of the Table given in § 61 shows, that although the series of values of $H \sqrt{H} - h \sqrt{h}$ does not differ much from the series of the corresponding discharges, yet that it follows them less closely than those of $H \sqrt{H}$. Thus in this important point, the formula in the last column is not so well established as that which precedes it. It is also of more difficult application, containing an additional term, and one whose exact determination is a matter of very great difficulty.

66. *Observations on the Formula in* § 56.—This formula, which involves a term that is a function of the velocity with which the water flowing in the canal arrives at the overfall, is well founded and of practical utility.

It is evident that in the case of a high velocity, in which the flow takes place both from the charge H and from an *initial velocity*, w, taken at the surface, it is necessary to add to the charge a term depending on this velocity, which leads to the equation, § 56,

$$Q = 5.35 \; m'l \times H \sqrt{H + 0.035 \; w^2}.$$

The experiments of M. Castel give the values of m' the coefficient. In these experiments the velocity w of the surface of the water in the canal has not been actually measured, but it can be determined from the mean velocity, which is equal to the discharge Q, divided by the section of the running water in the channel of supply, which in this case is $L \times (H + a)$, representing by L the width of the channel of supply, and by a the height of the sill of the overfall above the bottom of the channel; and as

$$v \times \{L\,(H + a)\} = Q. \quad \text{We have } v = \frac{Q}{L\,(H + a)}.$$

It will hereafter be shown that the velocity of water at the surface is, on an average, a fourth part higher than the mean velocity; so that we have—

$$w = 1.25 \times \frac{Q}{L\,(H + a)}.$$

Even with this value of w—which is the highest we may assume—the coefficient m' differs from the coefficient m in the common formula only when the velocity in the

channel is great enough to occasion the term $0.035\ w^2$
—which is that which makes the difference between the
two formulæ—to have a value comparable to that of H.
As it will be in most cases very small, and as it is under
the radical sign, it will only influence the value of m' by
half its amount relatively to H; for example, if it is the
2, 4, or 6 hundredths of H, the coefficients, *ceteris paribus*, will only differ by the 1, 2, or 3 hundredths. In these
three cases the section of the sheet of water at the overfall, or $l \times H$, is found to be respectively equal to 0.1724,
0.244, and 0.3, of the section of the canal of supply, or of
$L(H + a)$, whence may be deduced the conclusion of which
we have already made use (§§ 61, 64), that when the former of these two sections is less than the $\frac{1}{5}$th part of the
latter, the coefficients m and m' will be the same within
a hundredth part nearly. Such has been the case with
the overfalls used by M. Castel, whilst their width has
been less than the half of that of the channel. When
it was greater, the term $0.035\ w^2$ has had more effect,
and the differences became larger. But the use of this
term is far from having brought to equality the coefficients m and m' for various widths of overfall; it has
not reduced even by half the differences which occur in
the values of m; and neither the expression—

$$Q = 5.35\ m'l \times H \sqrt{H \times 0.035\ w^2},$$

nor that of

$$Q = 5.35\ ml \times H \sqrt{H}.$$

can be employed with a constant coefficient, except in
the case of a width of overfall equal to that of the canal
of supply.

In order to obtain the coefficient in this case, M.
Castel dammed up the canal of 2.427 feet by means of
barriers of copper, whose height was decreased succes-

sively from 0.738 feet to 0.104 feet : and he has thus obtained the coefficients in this Table :—

TABLE *showing the Variation of the Coefficient in the Formula* § 56, *due to circumstances affecting the Velocity of Approach.*

Height of sill of Dam above bottom of Channel.	Coefficients m', the Charge being			
	0.26 Feet.	0.19 Feet.	0.16 Feet.	0.13 Feet.
Feet.				
0.738	0.651	0.655	0.657	0.660 ⎫
0.557	0.640	0.647	0.650	0.654 ⎪ mean of all
0.426	0.650	0.649	0.652	0.656 ⎬ 0.650
0.305	0.635	0.642	0.646	0.650 ⎪
0.246	0.647	0.652	0.655	0.660 ⎭
0.134	0.667	0.664	0.665	0.668
0.104	0.676	0.676	0.676	0.680

Those of the five first barriers give nearly the same coefficient; and although they do not present the same regularity which we had in ordinary overfalls, we may assume 0.650 as the mean.

As to the last two barriers of 0.134 feet and 0.104 feet, they are in a distinct class, they were very low, and in them the charges very much exceeded the height of the sill above the bottom; so that the case was nearly as much one of water flowing in an unobstructed channel, as of passing over the sill of a waste-board. It may be remarked, also, that the near equality between the coefficients for each barrier in the horizontal line of coefficients speaks strongly in confirmation of the formula which has determined them.

The experiments upon the canal of 1.180 ft., with its barrier of 0.557 ft., have indicated coefficients whose mean was 0.654. Taking, then, the mean between this and 0.650, that is, 0.652, and observing that 5.35 × 0.652 = 3.488, we shall have, finally, for the discharge, with a velocity of approach equal to w feet per second on the surface—

$$Q = 3.488 \, LH \sqrt{H + 0.035 \, w^2}.$$

The velocity w is to be determined by direct observation.

67. *Overfalls, with Channels attached.*—We frequently find channels are adapted to overfalls : they may be considered as the prolongation of their horizontal and vertical edges. The water discharged is now confined, and suffers a resistance from the friction of the bottom and sides, which retards the motion, and this retardation, re-acting on the water which arrives at the overfall, diminishes the discharge. The following experiments by MM. Poncelet and Lesbros exhibit the effects of this resistance. The additional channel was always $3^m = 9.84$ feet long, and of the same width as the overfall, $0^m.20 = 0.656$ feet, and adjusted so as to be horizontal :—

TABLE *showing the Effect upon the " Coefficients" produced by Channels added externally to an Overfall.*

Charge.	Coefficient.		Loss per cent.
	Without added Channel.	With added Channel.	
Feet.			
0.675	0.582	0.479	18
0.475	0.590	0.471	20
0.337	0.591	0.457	23
0.196	0.599	0.425	29
0.147	0.609	0.407	33
0.091	0.622	0.340	45
Mean, 0.430.			

The amount of diminution in the discharge from *over-falls* with the channel attached has therefore been so much the less, as the charge has been greater. From this we may infer, that with charges of 3 or 4 feet and greater, such as are often in operation at the head of large feeders and water-courses, the diminution of discharge due to the resistance of the bottom and sides of the channel is inconsiderable. With *orifices* the same experimenters arrived at results analogous to those of overfalls. They applied the additional channel, 9.84 feet long by 0.656 feet wide, mentioned above, to the exterior of the orifices from which had been derived the Table § 21, where it is mentioned that the orifices were all 0.656 feet wide; and from numerous experiments it was deduced that when the charges, measured from the centre of the orifice, were from 2 to 2⅓ times greater than the height of the orifice itself, the channel attached had no decided influence upon the discharge: it was the same in amount whether this was or was not present; but with very small charges it diminished the discharge even a fourth or more.

Further investigations as to the effect of inclining the channel attached to the orifice were undertaken. When the slope was 1 in 100, or 34′, the coefficients were found the same as when the channel was horizontal, but at 1 in 10, or 5° 54′, they were increased 3 or 4 per cent. Castel also experimented on overfalls, and on his canal of supply, 2.247 feet wide, he placed an overfall of 0.656 ft. wide, with a channel attached 0.669 ft. long, inclined 1 in 13.3 or 4° 18′:—

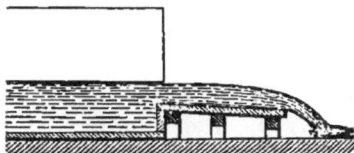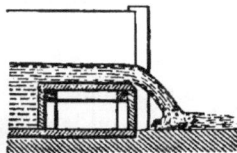

Fig. 25.   Fig. 26.

TABLE *showing the Result of Castel's Experiments on Channels added externally to Overfalls.*

Charge.	Coefficients.
Feet.	
0.364	0.526
0.311	0.527
0.249	0.527
0.196	0.528
0.164	0.530

The coefficients were obtained with the formula,

$$Q = 5.35 \, mlH \sqrt{H}.$$

They vary very slightly, although the charges were more than doubled. The mean is 0.527, and would probably have been 0.53 if the inclination had been 1 in 12, which is common in practice. With the simple overfall the co-efficient was 0.60; so that the additional channel diminished the discharge about 12 per cent.

68. A particular case yet remains to be considered— namely, that of the *demi-deversoirs* or *deversoirs incomplets,* as Dubuat has called them, or drowned weirs in English writers, so called when the tail water has risen above the level of the sill, Fig. 27. Dubuat divides the height of the water above the sill into two parts, A*b* and *b*C. In

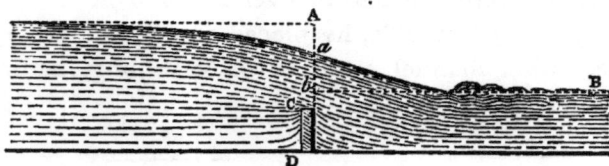

Fig. 27.

the former, the flow takes place as in an ordinary overfall, in which A*b* ($= H$) is the charge; so that the discharge (§ 66) is expressed by $Q = 3.488\ lH$ $\sqrt{H + 0.035\ w^2}$. In the second portion it may be assumed that the discharge is the same as in a rectangular orifice, of which bC is the height, and the charge equal to Ab the difference of level between the upper and lower surface of the water. bC is the height of this latter surface above the sill of the overfall, and if we call the height bD of the surface above the bottom of the canal, a, and the height CD of the sill above the same point, b, it will be equal to $a - b$. To the charge Ab or H is to be added, as in the case of closed orifices, the height due to the velocity u of the water in the canal, and the velocity of issue will be—

$$V = \sqrt{2g\ (H + 0.01553\ u^2)} = \sqrt{2g\ (H + 0.01\ w^2)},$$

For since the surface velocity is generally greater than the mean by one-fourth, that is $w = 1.25\ u$, we have, squaring, $w^2 = 1.5625\ u^2$, and dividing $\dfrac{w^2}{1.5625} = u^2$; but at page 62, § 56, it is shown that the additional head, h, due to antecedent velocity, is expressed by $0.01553\ u^2$, so that $h = \dfrac{0.01553\ w^2}{1.5625} = 0.01\ w^2$, as in the above equation. Thus we shall have for the discharge of this part (§ 29)—

$$Q = 4.96\ l\ .\ (a - b)\ \sqrt{H + 0.01\ w^2};$$

adding the two partial discharges, and putting Q for the total discharge—

$$Q = 3.488\ lH\ \sqrt{H + 0.035\ w^2} + 4.96\ l\ (a - b)\ \sqrt{H + 0.01\ w^2}.$$

69. *Arrangements preliminary to guaging.*—Permanent weirs constructed completely across the bed of a river may sometimes give the means of measuring the

discharge; but it is in such cases necessary that the crown of the weir should have a well-defined edge, so that the water which flows over it may fall freely, and without meeting any check from the reaction of the body of water already passed over, as in the experiments with added channels (§ 67). It is but seldom that they are so constructed; however, we may, without great expense, adapt a weir with the usual rounded crest to purposes of guaging by raising on the crest some temporary structure that shall have the necessary well-defined edge, and of a sufficient height, so that the discharge may not suffer from any such reaction.

We must so regulate the length of this apparatus that the depth of the water, *H*, flowing over may not be less relatively than the fourth part of the depth of the river as it approaches the weir. In such cases the discharge will be given by the formula $Q = 3.558 \, LH\sqrt{H}$ (§ 64), *L* being the length of the tempory crest or edge. In case *H* should exceed the fourth part of the depth of the current of the stream, we must use the formula in § 66, $Q = 3.488 \, LH\sqrt{H + 0.035 \, w^2}$, in which *w* is the velocity taken at the surface.

70. If the mode of gauging by overfalls be but seldom applicable to large rivers, it is, on the other hand, the most suitable for small streams and water-courses. We divide these into two cases: first, those in which the quantity of water discharged is at or under about 40 cubic feet per second; and secondly, those discharging more than that quantity.

A spot must be chosen where an overfall can readily be established; and in order that $Q = 3.21 \times l \times H\sqrt{H}$ may be safely applied, it should have a length, *l*, greater absolutely than 0.3 feet, but less relatively than the third part of the width of the bed of the stream, and so dis-

posed that we may have a charge, *H*, greater than 0.1968
ft. ; all being, moreover, subject to the condition that the
area *l* × *H* be not greater than the fifth part of the transverse section of the current immediately above the overfall ; then, without fearing any error greater than the
hundredth part, we may apply the formula

$$Q = 3.21 \, lH \, \sqrt{H} \; (\S \, 64)$$

Secondly, if the quantity discharged should exceed
40 cubic feet per second, we must pond up the water by
a dam extending from bank to bank, and at each extremity place vertical side-boards, so that the opening
traversed by the water may be rectangular, the crest
being truly horizontal, and using either of the formulæ
mentioned above, according to the conditions specified—
that is, according as there is velocity of approach or not.
The two examples following will point out the manner
of proceeding, and furnish an opportunity of adding some
practical details, elucidating what has been laid down in
§ 63 to § 66.

71. First. Let us suppose it necessary to gauge the discharge of a small river or water-course ; we must search
for a part suitable for the construction of an overfall.
This will probably be found at a point where the bed has
become contracted, and the banks are somewhat steep,
and immediately below a wide part of the stream; at
such locality the width at the water surface is found to be,
suppose, 11.808 feet, and the greatest depth 2.62 feet.
After a preliminary examination of the transverse section, and of the surface velocity, measured by means of
some light body thrown into the current, we find, approximately, multiplying the assumed section by the
velocity, that the stream is discharging nearly 35 cubic
feet per second.

Since the width is 11.808 feet, we may give 3.936 feet
to that of the overfall for gauging (that is, 11.808 ÷ 3) ;

the charge H will then be about 1.97 feet, for the formula $Q = 3.21\ lH\sqrt{H}$ gives—

$$H^{\frac{3}{2}} = \frac{Q}{3.21 \times l} \text{ or } H = \sqrt[3]{\left(\frac{Q}{3.21 \times l}\right)^2}$$

which, for the assumed discharge, gives in this case,

$$H = \sqrt[3]{\left(\frac{35}{1.321 \times 3.936}\right)^2} = \sqrt[3]{(2.7703)^2} = 1.9725 \text{ ft.}$$

From this preliminary inspection we maӯ construct a suitable partition of planks about 0.1 feet thick on the upper edge, and of such figure as nearly to fit the sides and bottom of the water-course. It must be carefully staunched, being sunk into the bottom and sides, and puddled on the up-stream face. From out of its upper edge we must cut an opening 3.936 ft. long and 2.132 ft. deep, so that its sill being 0.488 feet above the bottom of the bed (2.62 − 2.132), the water may flow off freely. The section of the presumed discharge (3.936 feet × 1.969 = 7.746 square feet) being not the fifth nor even the seventh part of the transverse section of the river, which exceeds 55 square feet, all the conditions for the application of the formula $Q = 3.21\ lH\sqrt{H}$ are present. Everything being duly prepared for the gauging of the water —such as, all leakage having been stopped, and the current restored to its ordinary and uniform flow, we proceed to measure H by stretching a cord across the opening, whose ends are fastened to points in the sides, marked at the level of still water (deduction being made for capillarity), and about a foot from the vertical side of the overfall. The depth of the sill below this line at the centre of the opening is carefully measured, and found to be 2.01 ft., and the length also, intended to be 3.936 feet, is found to be 3.92 feet. The discharge is, therefore,

$$Q = 3.21 \times 3.92 \times 2.01 \sqrt{2.01} = 35.86 \text{ cubic feet per second.}$$

72. Secondly. A question of law requires that the exact quantity of water flowing down a stream when the surface is level with the top of a certain fixed mark be determined. The gauging must evidently be effected by a dam across the course. About 170 feet above the mark a temporary dam is placed, at a part where, from its regular width and inclination, the river-bed is suitable, having, when the water is at the height above named, a breadth of 64.94 feet, and a mean depth of 4.1 feet; the overfall being a plank well squared, and 0.1312 feet thick, the upper edge being placed truly horizontal, and 0.656 feet above the bench-mark. At each extremity a vertical piece is raised, so that the length of the overfall is 63.6 feet; close to the vertical pieces two others are placed, on which a scale is drawn whose zero is the upper edge of the plank.

These arrangements being made, it is only necessary to observe when the surface of the water down stream is level with the fixed mark, and then read the height of the water upon each scale. This last has been found to be 2.34 feet. As this height is nearly the half of that of the temporary dam (4.1 + 0.656 = 4.756), we cannot apply with confidence the formula 3.558 $LH \sqrt{H}$, § 64, p. 76, and we must use, § 66—

$$Q = 3.488 \, LH \, \sqrt{H + 0.035 \, w^2}.$$

To obtain the value of w, the velocity of the surface of the current on arriving at the overfall, we must take a distance of, say, 165 feet on each bank, above the point where the surface of the current begins to curve towards the overfall; mark these points, and, about 60 or 70 feet above them, place in the current some floating body of the same specific gravity as water, and mark carefully

the time which it may take to flow along the 165 feet :
a mean of six observations gave 48½ seconds ; whence—

$$w \text{ is} = \frac{165}{48.5} = 3.4 \text{ feet per second, and } 0.0349\, w^2 = 0.40344.$$

Hence—

$$Q = 3.488 \times 63.6 \times 2.34 \, \sqrt{2.34 + 0.40344} = 859.78 \text{ cubic feet,}$$

the formula $3.558\, LH\sqrt{H}$ would give 810.63 cubic feet.
Thus we may certify, that at the given height of surface
the river discharges about 850 cubic feet of water per
second.

73. Experiments on weirs on a large scale have been
undertaken by Mr. T. E. Blackwell. The first set, made at
a large sidepond on the Kennet and Avon canal, consisted
of a series of 243 experiments, made on overfalls of 3, 6,
and 10 feet in length, with heads from 0.0833 ft. to
1.166 ft., that is, from 1 to 14 inches, and with the varying circumstances of having for the overfall bar,—first,
a thin plate ; secondly, a plank 0.166 feet = 2 inches
thick, square on the top ; and thirdly, a crest or channel
attached externally, 3 ft. in length. The thin plate
was a piece of iron fender plate, barely 0.0052 ft. = $\frac{1}{16}$
inch thick ; and the broad crest was an apron formed of
deal boards 3 ft. long, roughly planed over, and fastened
on to the outer edge of the vertical overfall plank, so as
form an uninterrupted continuation of it, the object being
to approximate towards the case of wide-crested weirs
in good repair, such as may be found in actual use in
rivers, &c. : the position of this planking was, in some
of the experiments, horizontal, and in others sloped at 1
in 18 and 1 in 12.

74. The mean coefficient with a thin plate of iron
was 0.649, and with this form of overfall the highest
coefficient, 0.808, occurred with one inch depth of water,

the lowest 0.529 with 9 inches depth; the intermediate
charges giving coefficients decreasing nearly uniformly
from the highest.

The mean coefficient, with an overfall plank 2 inches
thick, was 0.564, but in this case the lowest coefficient is
found with one inch charge, namely, 0.453, that at 9
inches depth being 0.575, contradicting the results obtained with the thin plate. And not only in this point,
but they differed also as regards the effect of different
lengths, for, with 3 ft. length of crest, the former gave a
less coefficient, namely, 0.631, than with 10 feet length,
which was 0.667; whereas, with the 2 inch plank for
overfall, the 3 feet length gave 0.570, and the 10 feet
length, 0.556.

The effect of wing-boards converging towards the
crest was shown to be advantageous, for, with 10 feet
length of crest, and 2 inch thickness of overfall, the coefficient was, with wing-boards, 0.688, and without them,
0.564. The boards converged at an angle of 64°, and
gave an increased discharge of nearly 20 per cent.

The planking of 3 feet length attached to the outside of the crest, and having a slight slope of 1 in 12 and
1 in 18, reduced the coefficient to 0.508 at a mean. When
placed level the mean was only 0.473.

75. The longitudinal section, Fig. 28, of the channel

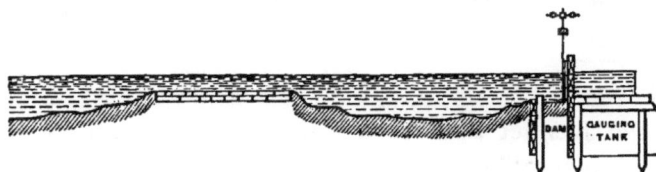

Fig. 28.

leading up to the overfall, shows that at some distance
above, the depth of the water was reduced by a submerged
course of masonry to about 18 inches, the overfall also

was placed at the outer edge of the dam to obtain the requisite free fall, so that the depth immediately in front was only 2 feet; the intermediate part being about 3 feet deep. From measurements of depth taken at still water, and corresponding depths of the sheet of water flowing off, it appeared that some degree of resistance opposed the motion of the water up to the overfall.

76. In the second set of experiments—those at Chew Magna—we have an area of reservoir of 5717 square feet kept constantly full by a pipe 2 feet diameter, discharging from a head of 19 feet. The distance between the mouth of this pipe and the overfall was only about 100 feet; the water must, therefore, have retained some of its velocity on approaching the overfall; and indeed, with charges above 0.417 or 0.5 feet, this was perceptible to the eye, but could not, the author states, be accurately determined, from the peculiar form of the reservoir. The results, however, show that this influence must have been considerable, and that the effect of water approaching an overfall with an initial velocity is an element which should never be disregarded. The longitudinal section, Fig. 29, as compared with the first set, Fig. 28,

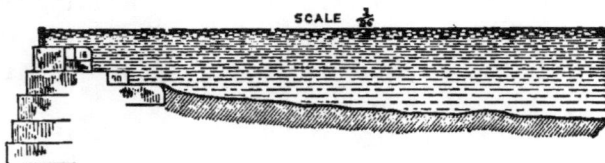

Fig. 29.

at the Kennet and Avon Canal reservoir, seems also more favourable to the free approach of the water. The overfall had wings at an angle of 45°, well adapted for facilitating the discharge. The overfall bar was a cast-iron plate, 0.166 feet thick, with a square top. The general circumstances attending these, the second set of experiments, make the discharges given by them analogous to the case of a weir in a river, or in a running stream ; and in this view they have great value when carefully applied.

In this second set the length of crest was in all the experiments 10 ft., and the overfall, a cast-iron plate, 2 inches thick and rectangular on the top. From these the mean coefficient deduced was 0.723. The head of water flowing over ranged from 1 inch to 9 inches, and the coefficient increased in proportion with considerable regularity—namely, from 0.591, with 1 inch depth, to 0.781, at 9 inches.

Whatever anomalies exist in these experiments of Mr. Blackwell, we may certainly claim that, where most consistent among themselves, they confirm the results of Castel, &c., and, being on a vastly larger scale, afford an answer to those who would undervalue the latter as being too small.

77. These several sets of experiments, and those of J. B. Francis, § 80, are probably, as to length of overfall, and charges some of which exceed 0.75 feet, the largest that have yet been recorded. To compare them fully with those of D'Aubuisson, it would, perhaps, be necessary to have had a greater number of widths below that of 10 feet than 6 and 3 feet, but with these only, as published, the comparison tends, in some degree, to confirm the coefficients given by D'Aubuisson in §§ 64, 66. Let us, for this comparison, recapitulate what has been there laid down. First, from the Tables, § 60, and the remarks, § 64, it appears that, with an overfall whose length is one-third or less of the channel of supply, we should use 0.60 as the multiplier or coefficient to reduce the expression—

$$\frac{2}{3} \, l \times H \, \sqrt{2gH,}$$

so as to give the true discharge per second (p. 76). Hence

$$Q = 3.21 \; l \times H\sqrt{H}.$$

Secondly, when the overfall is of a length equal to that of the width of the channel of supply, we may use 0.665, provided also that the head of water be less than one-third of the height of the dam above the bottom of the channel of supply, giving—

$$Q = 3.558 \; LH\sqrt{H}.$$

Thirdly, we have coefficients which decrease from 0.662 to 0.595 in proportion as the length of the overfall ranges from the full width of the channel to the fourth part of it. Lastly (§ 64, pp. 76, 77), we have 0.652 to be employed with overfalls whose length is the full width of the channel of supply; and when also the initial velocity of the water in this channel is represented in the formula by the quantity added to H in the factor—

$$\sqrt{H + 0.035 \; w^2}, \text{ giving } Q = 3.488 \; LH \sqrt{H + 0.035 \; w^2}.$$

Fourthly, we have, in the case of the additional channels or broad crests, the coefficient 0.527 to be used for m in formula—

$$Q = 5.35 \; m \times l \times H \sqrt{H}, \text{ giving } Q = 2.8355 \; lH\sqrt{H}.$$

78. In Mr. Blackwell's experiments we can only use for this proposed comparison those in which the overfall was constructed of plate-iron 0.0052 feet thick; and secondly, those having the crests 3 feet broad attached below the edge; since D'Aubuisson has only recorded experiments made with a thin plate of copper for the overfall, and those which had a channel attached below the edge. Now, if we consider the 10 feet overfalls of Mr. Blackwell as being analogous to those in D'Aubuisson, in which the length of the overfall is equal to the full width of the channel of supply, we find that the ave-

rage coefficient, the overfall being of thin plate-iron and 10 feet long, is—

According to T. E. Blackwell, . . . ˙. . 0.667

      ,,   D'Aubuisson (§ 64. second case), 0.665.

The average of the 3 feet overfall (less than one-third of the channel of supply), constructed of plate-iron 0.0052 feet thick, according to Blackwell, is 0.631, in the "first case;" in § 64 we find 0.600 as being used by D'Aubuisson in analogous circumstances. It may, however, be observed, that the charges in Mr. Blackwell's experiments, which give 0.631, not going higher than 0.5 feet, and the coefficients decreasing up to that head, make it probable that they would have decreased much lower had the experiments been continued, and so reduced the coefficient 0.631 to a value nearer to 0.600.

Again, in the experiments with the added channels or broad crests, we find the average of Mr. Blackwell, when the crest is horizontal, to be 0.473. The average of those in D'Aubuisson, § 67, p. 82, is 0.430, but particular experiments give a closer agreement in the coefficients: for instance, if we take out that derived from the charge 0.337 feet in this last Table, the coefficient is 0.457; and under the nearly equal charge of 0.333, we have the same identical coefficient 0.457 as the mean of the two given for the 3 feet and 6 feet overfalls. Castel's experiments for overfalls, with channels attached, sloping 1 in 13.3, give at the mean 0.527; the mean of those sloping 1 in 12, is 0.508. In these also, if we take out the particular heads of 0.164 feet in the former, and 0.166 in the latter, we have the respective coefficients 0.530 and 0.532.

79. The overfalls having the sill or bar, in the first set of experiments of plank, and in the second set at Chew Magna, of cast-iron, each 0.166 feet thick, and with square top edges, represent a very common structure for waste weirs, tumbling-bays, &c., on artificial canals and

feeders. The average of the first set, with charges from 0.083 to 1.166, and 10 feet length of overfall, gives 0.556; that of the second set is as high as 0.723.

The plan and longitudinal section of the channel of approach is evidently more favourable in the latter case than the former; and the velocity of the approaching water must also have been considerable, from the circumstances mentioned in § 76. The very low coefficient 0.556 is not, however, readily to be accounted for; nor is it easy to assign a reason why, in the first set, the change from a thickness of 0.0052 feet to 0.166 feet in the overfall bar should lower the coefficients to such an extent, every other circumstance being apparently the same.

If we look to the coefficients of particular experiments we also find discrepancies, as, for instance—

Thickness of crest of Overfall.		Charge.		Coefficient.
ft.	in.	ft.	in.	
0.0052	= $\frac{1}{16}$	0.083	= 1	0.808
0.166	= 2	0.083		0.435
0.0052		0.75	= 9	0.529
0.166		0.75		0.558

Length 10 ft. {

The overfall of $\frac{1}{16}$th inch thick, which had, compared with that of 2 inches, the nearly double coefficient—the charge being one inch—has, when the charge is 9 inches, one somewhat lower. Again, with the overfall 3 feet long and charge of 6 inches, the coefficient—namely, 0.592, is the same with each of these thicknesses of crest.

However unaccountable the above discrepancies, Blackwell's experiments, in other parts, are consistent, and confirm those of Castel, &c. (§ 78), which is important, as the volume discharged was thirteen times larger than in the latter.

80. The following Table, compiled from various sources, exhibits at one view the results of different experimenters.

Overfall, 0.5 feet long.—SMEATON AND BRINDLEY.

Heads,			.083	.1042	.1146	.1354	.1927	.2604	.4166	.4687	.5417	
Coeffs.			.713	.681	.654	.638	.636	.602	.609	.571	.633	

Overfall, 1.533 feet long.—DU BUAT.

Heads,						.1482		.2666	.3887		.5627	
Coeffs.						.648		.624	.627		.630	

Overfall, 0.656 feet long.—D'AUBUISSON AND CASTEL.

Heads,			.098	.131	.164	.196	.262	.328	.393	459	.524	.590
Coeffs.			.632	.624	.620	.617	.616	.617	.620	.624	.628	.633

Overfall, 0.6458 feet long.—PONCELET AND LESBROS.

Heads,	.033	.066	.099	.1332	.1998	.2664	.333	.5	.666	.75		
Coeffs.	.636	.625	.618	.611	.601	.595	.592	.590	.585	.577		

First Set.—Overfalls 3 feet and 10 feet long.—SIMPSON AND BLACKWELL.

Heads,			.083	.166	.25	.33	.416	.50	.583	.666	.75	
Coeffs.			.742	.738	.636	.635	.625	.592		.580	.529	

Second Set.—Overfall 10 feet long.—SIMPSON AND BLACKWELL.

Heads,			.083	.166	.25	.33	.416	.50	.583	.666	.75	
Coeffs.			.608	.682	.725	.745	.780	.749	.772	.802	.781	

Overfall, 10 feet long.—J. B. FRANCIS.—LOWELL EXPERIMENTS.

Heads,			0.62	0.65	0.80	0.83	0.98	1.00	1.06	1.25	1.56
Coeffs.			0.622	0.622	0.623	0.625	0.625	0·622	0.627	0.623	0.62

The total volume of water which passed into the measuring tank in the Lowell experiments was between 11,000 and 12,000 cb. feet. In every case except the last three we may perceive that the coefficient decreases as the

H

charge increases. Another exception may be found by referring back to the experiments of the first set,—overfall being a plank 0.166 feet thick,—in which the coefficients increase with the increase of the charges; the lengths being 3 feet, 6 feet, and 10 feet; and with the lengths of 6 feet and 10 feet attaining a maximum value at the charge of 0.583 feet from which they slightly increase.

Experiments on the smaller differ from those on the greater "heads" in this, that they can be, and generally are, continued for a longer period of time: a measuring tank of a definite capacity being always part of an experimental apparatus, the smaller discharges may be much prolonged, and thus all errors, such as in the noting the commencement and end of an experiment, are relatively diminished.

81. *Method of determining the Coefficient from Experiments.* — Smeaton's experiments were conducted by making observations upon the time in which a vessel of 20 cubic feet capacity was filled by the water flowing over a notch 0.5 feet long, and with the different charges given in the Table above.

Thus, with 0.1042 feet charge we should have, from the formula—

$$\frac{2}{3} lH\sqrt{2gH} = \frac{2}{3} \times 0.5 \times 0.1042 \times \sqrt{0.1042} \times 8.024 = 0.08995$$

cubic feet per second; but the experiment gives 20 cubic feet in 326 seconds, or (20 ÷ 326 =) 0.06135 cubic feet in one second, and—

$$0.08995 : 0.06135 :: 1 : 0.682 \left(= 1 \times \frac{0.06135}{0.08995}\right).$$

So also: Du Buat had a gauging reservoir to receive the discharged water, whose area was 108 square feet French; the water discharged by a notch (*reversoir*) 17 pouce 3 ligne long, with a head of 1 pouce 8 ligne, raised the surface of the reservoir above mentioned 1 pouce in three minutes: hence the discharge per second was 15552 × 5 ÷ 180 = 432 cubic pouce, and the formula gives—

$$Q = \frac{2}{3} \times 17.25 \times 1.666 \ \sqrt{1.666} \times \sqrt{2g};$$

and as with this unit $2g$ is equal to 724, we have $\sqrt{2g}$ equal to 26.907, the resulting value is 665.3 cubic pouce, and 665.3 : 432 :: 1 : 0.648; and in all experiments in general, the cubic quantity discharged in the observed time is to be reduced to the quantity per second, by dividing the former by the time expressed in seconds; and this, the actual discharge, being divided by the result of the formula expressed in numbers, gives the coefficient by which the formula must be affected to make its results coincide with actual experimental results.

Mr. T. E. Blackwell used in his first set a gauging tank of a capacity of 444.39 cubic feet: we find with an overfall 3 ft. long, and a head of 0.083′ feet, that in 757 seconds the discharge is 137.91 cubic feet, i. e. $\dfrac{137.91}{757} = 0.182$ cubic feet per second, now

$$\frac{2}{3} \times 3 \times 0.083 \ \sqrt{0.083} \times 8.024 = 0.3836 \text{ cubic feet,}$$

and 0.3836 : 0.182 :: 1 : 0.4744 = m. The average of three experiments, of which the above is one, gives m = 0.466 ; With an overfall of 10 feet long, and a head of 1 foot, we have the discharge equal to 442.29 cubic feet in 15.5 seconds, i. e. $\dfrac{442.29}{15.5} = 28.535$ cubic feet per second, and

$$\frac{2}{3} \times 10 \times 1 \ \sqrt{1} \times 8.024 = 53.493 \text{ ; therefore}$$

$$\frac{28.535}{53.493} = 0.5334 = m.$$

82. Mr. Beardmore, in his "Hydraulic Tables," has used the formula—

$$(a) \quad . \quad . \quad . \quad . \quad . \quad . \quad Q = 214 \sqrt{H^3},$$

in constructing the Table headed "Discharge of Weirs or Overfalls." This formula is very nearly identical with that in § 64, second case; for as this (a) gives the discharge, not for any length, l, but for one foot in length only, and per minute instead of per second, as all the formulæ given in the present work, we must, in order to compare them, divide (a) by 60, and multiply by l: hence, as $214 \div 60 = 3.566$, we have—

$$Q = 3.566 \, lH \sqrt{H} = \frac{2}{3} mlH \sqrt{2gH},$$

and consequently as—

$$\frac{2}{3} \times \sqrt{2g} \times m = 3.556, \text{ we have } m = \frac{3 \times 3.566}{2\sqrt{2g}} = 0.66654;$$

hence, we may write (a) thus,—

$$Q = \frac{2}{3} \times l \times 60 \times 3.566 \, H\sqrt{H},$$

which is adapted for any length l, and per second of time, and not per foot of length of overfall, and per minute.

This author also remarks, "That the constant 214 is liable to some variation under unfavourable circumstances: for instance, when the weir is formed of a number of short bays, divided by vertical beams, grooved for sliding down the horizontal waste-boards to regulate the surface-level of top water. In these cases, the water passing the edges assumes the *vena contracta* form in each bay, and, consequently, the total width, L, of the opening should be reduced to obtain the true quantity of water passing. These, and other causes which may render the observations liable to error, must be treated with judgment, according to circumstances." .... "The best way of gauging for the value of *H* in weirs is to have a post with a smooth head, level with the edge of the waste-board or sill : to be driven firmly in some part of the pond above the weir which has still water. A common rule can then be used for ascertaining the depth, or a gauge, showing at sight the depth of water passing over, may be nailed with its zero at the level of the sill of the weir. Among the conditions essential to a correct result are the absence of wind and current, a good thin-edged waste-board, the water having a free fall, and a weir not so long in proportion to the width above it as to wire-draw the stream ; for in this case the water will arrive at the weir with an initial velocity due to a fall, which is not estimated in the gauging, and the result will in all probability be too small, unless it be fully estimated for in the formula employed."

CHAPTER II.

FLOW OF WATER UNDER A VARIABLE HEAD.

83. *Flow of Water when the Level is variable upon one or both Faces of the Orifice of Discharge.*—When a reservoir, instead of being maintained constantly full, as we have supposed it to be hitherto, receives no supply, or receives less than it discharges through an orifice in the bottom, the surface of the fluid gradually descends, and the tank or reservoir is at length emptied. The laws of the discharge are in this case different from those which have been stated in the first chapter, and the questions to be resolved are of a different character.

The form of the vessels may be also, either prismatic—that is, of identical sections at every height of the surface—or having sides sloping at some known inclination.

84. *Ratio between the Velocities at the Orifice and in the Vessel.*—Let us suppose that the fluid contained in a prismatic vessel be divided into extremely small horizontal sections, and that they descend parallel to each other, the particles of the fluid in each of the sections must then have the same velocity. This is the hypothesis of the parallelism of the horizontal sections, admitted, and perhaps too much extended, by many hydraulicians.

Let v be the velocity of the particles in the vessel; V that which they have at the orifice; A the horizontal section of the reservoir or vessel containing the water; S, or rather mS, that of the orifice; m being the coefficient of contraction, the volume of water which flows out in the indefinitely small time τ will be expressed by $mSV\tau$.

During this same time the surface of the water descends by a quantity v_T, and the corresponding value of the volume of water is $A v_T = mSV_T$, or $v : V :: mS : A$, giving an example of that hydraulic axiom—namely, that the velocities are in the inverse ratio of the various transverse sections.

85. *Head due to the Velocity of the Water at its Point of Discharge.*—The velocity V of the issuing fluid does not now maintain the same constant rate. It is uniform only for a given instant; for, besides being due to the actual head at the given instant, the velocity V is a consequence of the velocity v acquired during the descent of the parallel sections above mentioned : the two velocities acting in the same direction, from above downwards, the result is equal to their sum. Thus, if $H' = \dfrac{V^2}{2g}$ be the height due to the velocity of the water at its point of discharge, H being always the actual head in the vessel, we shall have—

$$H' = H + \frac{v^2}{2g} = H + \frac{V^2}{2g} \cdot \frac{m^2 S^2}{A^2} = H + H' \frac{m^2 S^2}{A^2},$$

whence we have $H' = H \cdot \dfrac{A^2}{A^2 - m^2 S^2}$.

When mS is small compared with A, as is generally the case, $m^2 S^2$, with regard to A^2, may be neglected; so that $H' = H$, that is, the velocity of issue at any given instant is that due to the actual head at that same moment. In this chapter it is always assumed to be so, although the hypothesis of the parallelism of the horizontal sections, however admissible in their descent, does not hold good when they have arrived near the orifice, the circumstances of the movement of the molecules of the fluid become then very complicated, and are indeed entirely unknown.

86. *Nature of the Motion.*—Let M (Fig. 30) represent a vessel of water filled up to AB; let us divide the height from B to the orifice D into a great number of equal parts, Ba, ab, bc, &c. Suppose, then, that a body, P, were impelled from below upwards with a velocity such that it rises to the point H, PH being equal to DB, and let us divide PH into the same number of equal parts.

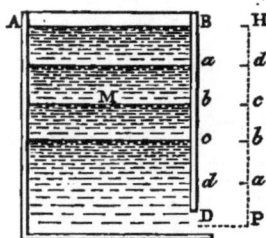

Fig. 30.

In proportion as the body rises, its velocity will diminish, in such a manner that when it shall have arrived successively at the points a', b', c', the velocities will be respectively $\sqrt{Ha'}$, $\sqrt{Hb'}$, $\sqrt{Hc'}$ . . . o, as is shown in works on the Elements of Mechanics. Recurring to the fluid contained in the vessel M, in proportion as it flows out, the surface AB is lowered; and when it shall have successively reached the points a, b, c, the respective velocities of the issuing water will be (§ 85) as $\sqrt{Da}$, $\sqrt{Db}$, $\sqrt{Dc}$ . . . o, or, according to the construction, as their equals $\sqrt{Ha'}$, $\sqrt{Hb'}$, $\sqrt{Hc'}$ . . . o; so that, in proportion as the vessel is emptied, the velocity of the discharge will decrease down to zero, following the same law as the velocity of the body impelled from below upwards, each being an example of an uniformly retarded motion; consequently, the discharge also will be governed by the same law.

It will be the same, also, in the descent of the surface of the water in the vessel, which will be uniformly retarded, its velocity being in a constant ratio to that at the orifice,—namely, as the section of the orifice to the area of the surface of the water.

87. *Volume discharged.*—According to the laws of an

uniformly retarded motion, when a body, starting with a certain velocity, loses it gradually until it is reduced to zero, it only describes one-half the space it would have traversed in the same time if it had moved uniformly with the velocity with which it commenced the motion. Now the volume of water which flows out from any vessel until it is all discharged may be regarded as a prism, whose base is the orifice, and height the space which the first issuing particles would describe, with a uniformly retarded motion identical with that by which the discharge takes place; but if the same particles had always preserved their initial velocity (which is that due to the primary charge), the space described in the same time, or the height of the prism, and, consequently, the volume of water discharged, would have been doubled. Hence this theorem :—*The volume of water which passes through an orifice at the bottom of a prismatic vessel, receiving no supply, and therefore becoming empty, is only one-half of that which would be given during the time of complete discharge, if the flow had taken place under a constant charge equal to the primary.*

88. *Time which is required to empty a vessel.*—Let H be this charge; A the horizontal section of the vessel; T the time which it may require to be completely discharged. The volume of water discharged during this time—that is to say, all the water the vessel contains (above the orifice)—is $A \times H$. The volume, according to the theorem above, which would have been discharged in that time under the constant charge H, would have been $2 (A \times H)$. This same volume, or the discharge during the time T, is also equal to $mST \sqrt{2gH}$.

We may use the Italic capital H instead of the Roman H, conventionally applied hitherto in this formula (§ 44), since the orifice is now supposed to be in the bottom of the vessel, and therefore H and H are identical. Equating these two values, we have—

$$2AH = mST\sqrt{2gH},$$

and solving for T, we have—

$$T = 2 \cdot \frac{AH}{mS\sqrt{2gH}},$$

and dividing above and below by $\sqrt{H}$, we have, finally—

$$T = 2 \times \frac{A\sqrt{H}}{mS\sqrt{2g}}.$$

If we represent by T' the time which the volume AH would take to flow out under the constant head H, we should have had (§ 14)—

$$AH = mST'\sqrt{2gH}, \text{ or } T' = \frac{A\sqrt{H}}{mS\sqrt{2g}},$$

consequently, $T = 2T'$; that is to say, *the time which a prismatic vessel takes to be completely discharged is double that in which the same volume would flow out, if the head had remained constantly the same as it was at the commencement of the discharge.*

89. *Time which the Surface of the Water takes to descend a given Depth.*—Let t be the time sought in which the level descends the given depth a: now the time in which the whole volume would be discharged is (§ 88)—

$$\frac{2A\sqrt{H}}{mS\sqrt{2g}};$$

the head at the commencement being H; and putting—

$$H - a = h$$

for the head at the end of t, we have the time in which

the volume hA would be entirely discharged equal to—

$$\frac{2A\sqrt{h}}{mS\sqrt{2g}}.$$

Now, the time t, that in which the surface descends a height equal to a, is evidently the difference between the two expressions given above, that is—

$$(a) \ \ldots \ t = \frac{2A\sqrt{H}}{mS\sqrt{2g}} - \frac{2A\sqrt{h}}{mS\sqrt{2g}} = \frac{2A}{mS\sqrt{2g}}(\sqrt{H}-\sqrt{h}).$$

90. *Volume discharged in a given time.*—The above expression for the time which the water requires to descend any given height, by a simple transformation, gives both the value of a, and also the volume of water discharged during the given time : thus we have from the equation (a)—

$$\frac{tmS\sqrt{2g}}{2A} = \sqrt{H} - \sqrt{h},$$

and—

$$\frac{tmS\sqrt{2g}}{2A} + \sqrt{h} = \sqrt{H};$$

hence,

$$\sqrt{h} = \sqrt{H} - \frac{tmS\sqrt{2g}}{2A},$$

squaring both sides of the former equation—

$$\left(\frac{tmS\sqrt{2g}}{2A}\right)^2 + 2\frac{tmS\sqrt{2g}}{2A}\cdot\sqrt{h} + h = H);$$

hence—

$$\frac{tmS\sqrt{2g}}{A} \times \left(\frac{tmS\sqrt{2g}}{4A} + \sqrt{h}\right) = H - h.$$

Hence, substituting for $\sqrt{h}$ its value given above, and multiplying both sides by A, we have the discharge Q' for the given time—

$$Q' = (H - h)\, A = tmS\sqrt{2g}\left(\sqrt{H} - \frac{tmS\sqrt{2g}}{4A}\right).$$

91. *Mean hydraulic Charge.*—A prismatic vessel receiving no supply, discharges through an orifice S, during T seconds, having at the commencement the head H, at the end h; required the mean hydraulic charge H', by which, *cæteris paribus*, the same quantity is discharged: we have (§ 14)—

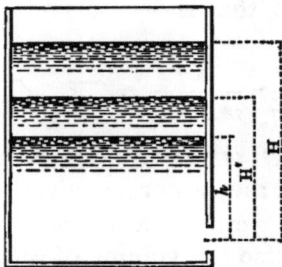

Fig. 31.

(b) . . . . . $Q' = mS\sqrt{2g}\, T\sqrt{H'} = (H-h)\, A$; also § 89 (a)—

$$T = \frac{2A}{mS\sqrt{2g}}\,(\sqrt{H} - \sqrt{h});$$

substituting this value of T in (b), we have

$$Q' = mS\sqrt{2g} \times \frac{2A}{mS\sqrt{2g}}\,(\sqrt{H} - \sqrt{h}) \times \sqrt{H'} = (H-h)\, A.$$

Clearing of fractions, and dividing, we have—

$$\sqrt{H'} = \frac{H - h}{2\,(\sqrt{H} - \sqrt{h})}, \text{ or } H' = \left(\frac{H - h}{2\,(\sqrt{H} - \sqrt{h})}\right)^2,$$

or

$$H' = \left(\frac{\sqrt{H} + \sqrt{h}}{2}\right)^2.$$

COR.—If $h = 0$, then $H' = \left(\dfrac{\sqrt{H}}{2}\right)^2 = \dfrac{H}{4}.$

92. *Case of a prismatic Basin receiving a constant*

Supply while discharging.—Let q be the volume received per second (less than that discharged), and x the space the water surface lowers in the time t: then dx will be its descent during the indefinitely small interval dt, and thus $A\,dx$ will express the volume flowing out during dt, if no supply entered; but as it receives q in one sec., and, therefore, $q\,dt$ in dt, the actual discharge will be $A\,dx + q\,dt$. From § 14 we thus have—

(a) . . . $A\,dx + q\,dt = mS\,dt\sqrt{2g\,(H-x)}$

putting $H - x = h$, and therefore $-dx = dh$, we have

(b) . . . $q\,dt - A\,dh = mS\,dt\sqrt{2g}\,\sqrt{h}$, which gives—

(c) . . . $dt = \dfrac{-A\,dh}{mS\sqrt{2g}\,\sqrt{h} - q}$.

In order to integrate this equation, we may put—

(d) . . . $mS\sqrt{2g}\,\sqrt{h} - q = y$, and thus—

$$dt = \frac{-A}{m^2 S^2 g}\left(dy + q\,\frac{dy}{y}\right),$$

the integral of which is—

$$t = \frac{-A}{m^2 S^2 g}\,(y + q \text{ hyp. log. } y) + C.$$

Determining the value of C for the commencement of the motion, when $t = 0$ and $x = 0$, and H also being equal to h, we have, substituting for y its value above,— C equal to—

$$\frac{2A}{(mS\sqrt{2g})^2}\,(mS\sqrt{2g}\,\sqrt{H} - q + q \text{ hyp. log } mS\sqrt{2g}\,\sqrt{H} - q).$$

Hence t is equal to—

$$\frac{2A}{(mS\sqrt{2g})^2}\left\{mS\sqrt{2g}\left(\sqrt{H}-\sqrt{h}\right)+q\,\text{hyp}\,\log\frac{mS\sqrt{2gH}-q}{mS\sqrt{2gh}-q}\right\}.$$

It is evident from this expression that when $q = 0$, that is, when no supply is flowing in, it becomes identical with that in § 89.

If we had to determine the height which the level of the water would descend in a given time, the question would be reduced to this other—namely, to find the charge h at the end of this time, and subtract it from H, the head at the commencement of the discharge. To obtain h we must substitute successive values of it; i. e. of $(H-x)$, in the equation given above, and thus tentatively determine that which satisfies the equation.

93. *Case when the Water is discharged over a Weir.*—In the case when the water issues from the basin by an over-fall, supposing that it receives no supply, we shall have, from what has been laid down in § 46 and § 55—

$$A\,dx = \frac{2}{3}ml\,(H-x)\,dt\,\sqrt{2g}\,\sqrt{H-x},$$

whence, by a method analogous to that which has been used above, we have—

$$t = \frac{3A}{ml\sqrt{2g}}\left(\frac{1}{\sqrt{h}}-\frac{1}{\sqrt{H}}\right).$$

94. *Reservoirs not being prismatic.*—We have hitherto considered only the particular case of prismatic basins or reservoirs: the determination of the time of discharge for any other form is much more complicated, and is even

impossible in most cases which present themselves. The fundamental equation is, however, always—

$$A dx = m S dt \sqrt{2g (H - x)},$$

from whence we have—

$$dt = \frac{A dx}{m S \sqrt{2g (H - x)}}.$$

But here A is variable, and we must, in order to integrate, express A in terms of x, which can only be effected when we know the law by which A decreases, and in the cases where the basin itself is a solid of revolution, whose generatrix is known. In every other case it will be necessary to proceed by approximations and by parts. To this end, we must divide the basin or reservoir into horizontal sections of small depth. Each of these may be taken as prismatic, and we can determine the time it takes to be discharged by the aid of the formula given above. The sum of these partial times will give the time that the surface of the water takes to descend a height equal to the sum of the heights of the prisms.

95. *Flow of Water when it is discharged from one reservoir into another.*

1st. In the case when the orifice is covered with water on both faces, the levels remaining constant, the quantity discharged is the same as if it had been into the air under a charge $H - h$, equal to the difference of the charges upon each face; thus we have, representing by Q the discharge per second,—

$$Q = m S \sqrt{2g (H - h)}.$$

2ndly. Let the level remain constant in the upper

basin, and the lower, of a given area, receive the discharge; required the time in which it reaches the level
of the upper basin or a given height. This problem is
the inverse of that in § 89, in which the surface of the
water descended with a uniformly retarded motion. In
the present case, the surface of the lower basin rises
with a uniformly retarded motion.

Let H represent the charge A C (Fig. 32) at
the commencement, and h the
charge AD at the end of the
time t, A the horizontal section of the vessel being filled,
and S and m as before,—we
shall have, for filling up to DE,

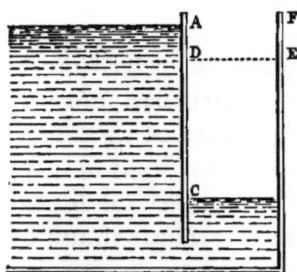

Fig. 32.

$$t = \frac{2A}{mS\sqrt{2g}} \left(\sqrt{H} - \sqrt{h} \right), \text{ and}$$

for filling up to AF,

$$T = \frac{2A}{mS\sqrt{2g}} \sqrt{H}.$$

These latter formulæ are of some importance : they
serve to determine the time in which canal locks, &c.,
may be filled, and to assign the area of sluice-way required to fill a certain volume in a given time.

96. *The Level of the Water being variable in each Vessel.*—We now come to the third case that can arise,
namely, when two reservoirs of different level communicate with each other, each being limited in area and receiving no supply, and thus one surface descends as the
other rises. Such is the case of the two basins K and L
(Fig. 33), communicating by a wide pipe EF, provided
with a sluice-door or cock at G. Before the opening of
this sluice-door the level of the water is at AB in the first
reservoir, and CO in the second. At the end of a certain

time after the opening of the communication it has descended to MN in the first, and has ascended to PQ in
the second. It is re-
quired to determine the
relation between these
two heights at a given
time, or, *vice versá*, from
the given difference in
the respective heights,
to determine the time

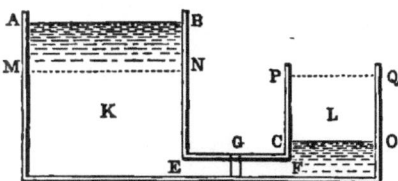

Fig. 33.

corresponding to a given discharge.

Let t equal the given time, BE = H, CF = h, NE = x,
PF = y, A = horizontal section of the first vessel, and B
that of the second, s = section of the pipe of communication: in the coefficient m we must include the resistance
of the water passing through this pipe. Whilst the water
has risen in the second basin by the quantity dy, during
the instant dt, it will have lowered in the other by dx;
and remembering that x diminishes while y and t increase, we have $A dx = - B dy$ and (§ 14),

(*a*) . . . . . $A dx = - ms \sqrt{2g(x-y)} \cdot dt,$

from whence—

(*b*) . . . . $dt = - \dfrac{A dx}{ms \sqrt{2g(x-y)}}.$

The first equation being integrated, remembering that
when $x = H$, $y = h$: we have—

(*c*) . . . . $Ax + By = HA + Bh$;

solving for y, we have—

$$y = \frac{A(H-x)}{B} + h;$$

and substituting this value of y in the preceding equation

I

(*b*), integrating, and observing that $H = x$ when $t = 0$, we have—

$$t = \frac{2A\sqrt{B}}{mS\sqrt{2g}(A+B)}\left\{\sqrt{B(H-h)} - \sqrt{(A+B)x - AH - Bh}\right\}$$

If it were required to find the time in which the two surfaces would be at the same level, we should have from (*c*) —

$$x = y = \frac{AH + Bh}{A + B};$$

and, this value of x being substituted in the above expression for t, will give—

$$(d) \quad . \quad . \quad . \quad . \quad t = \frac{2AB\sqrt{H-h}}{mS\sqrt{2g}(A+B)}.$$

From whence it is evident that for the same value of $(H - h)$ the time t is the same whether A be the horizontal section of the basin that lowers, and B that of the other, whose surface rises, or, *vice versá*, B that which falls, and A that which rises.

———— ————

---

97. THE following Rules, approximately true, may be found useful in every-day practice. It is important to know how they are derived, and thus be able to reproduce them, as no book may be at hand for reference, and the memory may fail. They all depend upon the volume and weight of water in relation to the weights and measures of the United Kingdom.

The statical pressure, i. e. of still water, in any pipe, or on the bottom of a tank, is qp, equal to 3lbs. per square inch for every 7 feet head. Thus a main laid across a valley is, let us suppose, at the lowest part, 130 feet below the surface from which it is supplied. From the rule $130 \div 7 = 18.57$ and $3 \times 18.57 = 55.7$ lbs. per square inch; this result is about the $\frac{1}{100}$th part too small, it should be 56.26 lbs.

If, on the other hand, we had a known pressure of water, suppose of 38.5 lbs. per square inch, to determine the vertical head in feet; by the Rule $38.5 \div 3 = 12.833$ and $7 \times 12.833 = 89.83$ ft. The exact result is 88.956 ft, so that when the pressure is given, the result is about the $\frac{1}{100}$th part too large.

Since a cubic inch of water weighs 252.458 grains, a column one foot high and one square inch in base weighs 3029.496 grains, which, divided by 7000 to reduce it to lbs. av., is $\frac{3029.496}{7000}$ = $\frac{3}{7}$th lbs. nearly, the exact fraction

being $\frac{3.0295}{7}$, or 3.03 lbs. for 7 feet, which is one per cent. greater.

In the same manner the longitudinal bursting pressure of water in a pipe per inch of length is found by multiplying the diameter in inches into the pressure per square inch—that is, $\frac{3}{7}$ × H ft. × D ins. Thus, if the diameter of a pipe be 26 inches, and H, as above, we have (using the exact result) 26 × 56.26 = 1462.8 lbs.

When computing the resistance against the plunger of a forcing pump in motion, it is usual to take half the height in feet for the pressure per square inch—that is, $\frac{3.5}{7}$ths of a lb. av. per ft. of height. Thus, to force water to the height of 47 feet we have 23½ lbs. per square inch resistance; this gives a fair allowance for friction, passing through valves, &c.

In pumping engines for mines it is useful to be able readily to compute the total weight of water in the vertical pipe at any lift, from that per yard or per fathom (= 6 feet). For this purpose;

Square the diameter in inches and the result is nearly equal to the lbs. per yard vertical, and for the fathom multiply this by 2; or per foot use 3 as a divisor.

Thus, in a pipe 13 inches in diameter and rising 40 fathoms we have 169 (= 13²) × 2 = 338 lbs. and 40 × 338 = 13520 lbs. The exact multiplier is 2.0454, giving a result a little more than 2¼ per cent. greater than the approximate rule. In all these the number of gallons is found by cutting off, from the number expressing the lbs. weight, one figure for decimals; thus in the length of 40 fathoms of the above pipe we have 1352 gallons, to which, adding 2¼ per cent., or 27¾, we have 1380 gallons. To prove the rule we have, putting d for the diameter in inches, and $\frac{11}{14}$ for $\frac{\pi}{4}$,

$$d^2 \times \frac{11 \times 36 \times 62.5}{14 \times 1728} = \text{weight in lbs. per yard,}$$

or as the multiplier of d^2 is equal to $\frac{24750}{24192}$, which being divided out equals 1.023, we have $d^2 \times 1.023 =$ weight in lbs. per yard.

The numbers 62.5 and 6.25, the former the approximate number of lbs. in a cubic foot, and the latter the number of gallons in the same, may, for facility of computation, be written $\frac{1000}{16}$, and $\frac{100}{16}$, the division by 4 × 4 being very easy. Hence these Rules. First. To change cb. ft. of water into lbs. Add three places and divide by 16. Thus the number of lbs. in 347.7 cb. feet is 347700 ÷ (4 × 4) = 21731.25 lbs.

And Secondly. To change cb. feet into gallons. Add two places and divide by 16.

The number of gallons in 893.47 cb. feet is

$$89347 \div (4 \times 4) = 5584.2 \text{ gallons.}$$

A distributing reservoir contains 21,450,000 cubic feet. Compute the number of days it would supply a town requiring 11,000,000 gallons per diem :

$$2,145,000,000 \div (4 \times 4) = 134,062,500,$$

and this, divided by the required supply of 11,000,000 gallons, gives 12 days.

Let it be required to calculate the number of cubic feet which an impounding reservoir should contain so as to be able to supply 25 gallons per diem to each person in a population of 85,000 for 200 days. For the above rate of supply we have this rule. Multiply the population by the assigned number of days and by four—

$$85000 \times 200 \times 25 \times \frac{4 \times 4}{100} = 85000 \times 200 \times 4 = 68 \text{ millions cb. ft.}$$

If the number of days assigned were 250, then, adding three cyphers, we have the result. Thus 126,000 inhabitants require a reservoir containing 126 millions of cb. ft. to give 25 gallons per diem for 250 days.

If, Thirdly, it were required to compute the number of cubic feet in, let us suppose, 337,489 gallons, multiply by $\frac{4 \times 4}{100}$, the reciprocal of 6.25, that is, cut off two places as decimals and multiply twice by 4;

$$3374.89 \times (4 \times 4) = 53998.24 \text{ cubic feet.}$$

A tank contains 1,457,965 gallons, to what number of cubic feet is that equal?

$$14579.65 \times (4 \times 4) = 233274.4 \text{ cubic feet.}$$

To change cubic feet per minute into gallons per diem. Multiply by 9000. For 1 cb. ft. $\times 60 \times 24 \times \frac{100}{16}$ $= 1 \times 15 \times 6 \times 100 = 9000$.

The well-known definition of Horse-power, that is, 33,000 lbs. raised one foot high in one minute, renders it easy to compute the power of any stream for mill work when once its discharge is known. It will be necessary to have a weight of about 45,000 lbs. of water per minute falling one foot to develope one horse-power at the point of application of the power, and this requires 720 cb. feet per minute, or 12 cb. feet per second, falling one foot.

98. *Questions solved by means of the Formula* $mS\sqrt{2gH} = Q$, *the Charge on the centre being represented by* H.—(I.) In order to obtain a comparative view of the effects resulting from the use of the different coefficients for the discharge through various orifices, given in §§ 18 to 43, to

which we first confine our attention, let us take a circular orifice of 0.25 ft. in diameter, the area S being therefore $0.25^2 \times 0.7854 = 0.04909$ sq. ft., and determine :— First, the discharge through it in some given time, as 40 minutes, with a constant charge of, suppose, 9 ft. above the centre of the orifice ; and, secondly, with the same orifice and charge, seek the different intervals of time required to discharge a given volume of water, as 1000 cubic feet. As the charge is so great compared with the diameter in the above data, we may use the formula (§ 14)—

$$Q' = mST\sqrt{2g}\sqrt{H},$$

in which H is the charge on the centre. In the first case mentioned above we calculate the value of—

$$ST\sqrt{2g}\sqrt{H},$$

which becomes 0.04909 sq. ft. × 40 min. × 60 × 8.024 $(= \sqrt{2g})$ × 3 $(= \sqrt{H}) = 2836.067$ cb. ft., and multiply it by the several values of m, as is done below. For the second case we have—

$$\frac{Q'}{S\sqrt{2g}\sqrt{H}} \times \frac{1}{m} = T \text{ seconds };$$

the value of the first factor of the left-hand side is—

$$\frac{1000}{0.04909 \times 8.024 \times 3} = \frac{1000}{1.1817} = 846.24 \text{ cb. ft.}$$

which must be divided also by the several values of m to obtain T, the time required to discharge the given quantity.

	Value of Q' in 40 min. $m \times 2836.067$.		Value of T, or 846.24 cb. ft. $\div m$		
			min.	sec.	
(1) m (§ 28) = 0.50	1418 cb. ft. . .		28	12	internal tube.
(2) m (§ 18) = 0.62	1758 . . .		22	45	thin plate.
(3) m (§ 34) = 0.82	2325 . . .		17	12	cylindrical adjutage.
(4) m (§ 40) = 0·95	.2694 . . .		14	51	conical converging adjutages.
(5) m (§ 18) = 1.00	2836 . . .		14	6	form of *vena contracta* and conl. converging.
(6) m (§ 43) = 1.46	4140 . . .		9	39	conical diverging adjutages.

(II.) Required the discharge in six minutes, through a rectangular sluice 3 ft. by 1 ft., the side 3 ft. long being horizontal, the depth to the sill from the surface being 7 ft., and m being equal to 0.62.

Here

$$0.62 \times 3 \text{ sq. ft.} \times 8.024 \sqrt{6.5} = Q,$$

and

$$\sqrt{6.5} = 2.5495 \text{ may be taken equal to } 2.55,$$

hence

$$Q = 38.06 \text{ cb. ft. per sec.,}$$

and

$$Q' = 6 \times 60 \times 38.06 = 13701.6 \text{ cubic feet.}$$

(III.) A reservoir having at full water a depth of 40 feet over the centre of the discharging sluice, whose area is 2 feet horizontal by 1.5 ft. vertical when fully opened :— Required the discharge at that depth, and also when the water has sunk to the heads, 30 ft., 20 ft., and 10 ft., the value of m being taken at 0.62 in each case,—we have—

$$S = 1.5 \times 2 = 3 \text{ sq. ft., and } \sqrt{40}, \sqrt{30}, \sqrt{20}, \text{ and } \sqrt{10},$$

being respectively 6.324, 5.477, 4.472, 3.162. We must multiply these numbers successively by—

$$0.62 \times 8.024 \times 3 = 14.92464,$$

which is the same in each. Hence, for 40 ft. head the discharge is 94.384 per sec.; for 30 ft., 81.742 cb. ft.; for 20 ft., 66.743 cb. ft.; and for 10 ft., 47.192, or the half of that for 40 ft.; 3.162 being necessarily half 6.324, as they are the roots of numbers in the ratio of 1 to 4. This question points out the fact that leakages of sluices in lock-gates, &c., increase far less rapidly than the head, being, in fact, as the square roots of the charges. (*Vide* Smeaton's Reports, vol. i., pp. 196–9).

(IV.) What is the discharge through a circular pipe 4 ft. diameter in the embankment of a reservoir, the head upon the centre being 90 ft., *m* being taken equal to 0.60 ? In this case—

$$S = (4)^2 \times 0.7854 = 12.5664 \text{ and } \sqrt{90} = 9.487,$$

hence

$$0.6 \times 12.5664 \times 8.024 \times 9.487 = 573.9 \text{ cb. ft. per sec.}$$

(V.) A rectangular sluice, sides 4 ft. horizontal and 3 ft. vertical, having a charge of 20 ft. on the centre, is raised 1.5 ft.: required the discharge per sec., and also when fully opened. We have the value of S in the first instance, one-half that in the second, but the heads to the centre of the orifice are 20.75 ft. and 20 ft. respectively; and assuming that $m = 0.62$ answers the particular circumstances of this orifice we have, first—

$$0.62 \times 8.024 \times 6 \times \sqrt{20.75} \ (= 4.5552) = 135.97 \text{ cb. ft. ;}$$

and secondly,

$$0.62 \times 8.024 \times 12 \times \sqrt{20} \ (= 4.472) = 266.97 \text{ cb. ft.}$$

The double of the former would be 271.94 cb. ft.

(VI.) In cities in which water is supplied at high pressure, and constant service, it is sometimes usual to give the water to manufactories and works through a very small orifice, perforated in a disc, which is closed up and secure from any possibility of unfair interference. Calculate the discharge through an orifice 0.0089 in. diameter for 24 hours, the head being 129 ft. and *m* equal to 0.62 ; we have—

$$\log. m + \log. S + \tfrac{1}{2} \log. (2g) + \tfrac{1}{2} \log. H + \log. 86400'' = \log. Q',$$

the log. of S being 2 log. 0·0089 + log. 0.7854,—we have thus—

$$Q' = 303.655 \text{ cb. ft.}$$

(VII.) Suppose the pressure on the mains to be measured by a head of 150 ft. of water, and the diameter of the orifice 0.02 ft. : required the quantity delivered in 24 hours, the coefficient of discharge being 0.62. The $\sqrt{150}$ being equal to 12.247, and—

$$S = (0.02)^2 \times 0.7854 = 0.00031416,$$

we have—

$$Q' = \begin{cases} T & \times & m & \times & S & \times & \sqrt{2g} & \times & \sqrt{H} \\ 24^h & \times\ 3600'' & \times\ 0.62 & \times\ 0.00031416 & \times\ 8.024 & \times\ 12.247 \\ & & & = 1653.7 \text{ cb. ft.} \end{cases}$$

(VIII.) What must be the diameter of the orifice to give 600 cb. ft. per diem, the head on the main being 100 feet ?

Here

$$S = \frac{600\ (= Q')}{24 \times 3600 \times 0.62 \times 8.024 \times 10} = \frac{600}{4298300} = 0.0001396 \text{ sq. ft.,}$$

and as $S = d^2 \times 0.7854$, we have —

$$d = \sqrt{\frac{0.0001396}{0.7854}} = \sqrt{0.0000177744} = 0.004216 \text{ ft.}$$

which is a little more than $\frac{1}{20}$th of an inch.

As the exact adjustment of this diameter would be nearly impossible, the process is somewhat tentative.

(IX.) In the sluices constructed in tidal harbours for scouring away at low water the silt that generally accumulates in them, we obtain examples on a very large scale of the discharge of water through orifices.

This simple remedy for a defect that had rendered nearly useless some of the most important tidal harbours on the coast of England, which had not the advantage of any sufficient natural streams to keep them open, was

introduced by the great Smeaton from his personal observation of the practice in the Low Countries (*vide* Reports,
vol. ii., p. 202–209). A bank thrown across some part,
covered at high tide, impounded the water allowed to
enter during the rise of the tide, and which at low water
is discharged very rapidly through sluices constructed in
this embankment, the sills of which are placed at low
water of springs, or as low as possible.

The practice subsequently fell into disrepute, as it
was found that the area of the back-water was itself soon
silted up ; but the same engineer adopted the simple and
efficient remedy of dividing the back-water into two separate areas by a second bank at or about perpendicular
to the first mentioned, and by occasionally using one of
these to cleanse the other, they were both, as well as the
harbour itself, kept clear. Ramsgate and Dover are
well-known examples (*vide* Smeaton's Reports and Sir
J. Rennie on Harbours); from which last-mentioned
work we take an example from the description of Hartle-
pool Harbour, on the coast of Durham.

Each sluice was 3 feet wide and 6.33 feet high, having
a charge estimated at 10 ft. on the average. From the
detailed plans of these works given by Sir J. Rennie, we
may consider the coefficient 0.600 applicable ; hence—

$$0.600 \times 3 \times 6.33 \times 8.024 \times \sqrt{10} = 289.14 \text{ cb. ft. per sec.}$$

is the discharge for each sluice ; and as it is also stated
that the total area of the scouring sluices was 366 sq. ft.,
of which 24 sq. ft. were given by four sluices, each 3 ft.
by 2, in the lock-gates, which communicated with the
back water or slake, we have 342 sq. ft. left for those
through the embankment ; and each of these being
$3 \times 6.33 = 19$ sq. ft., we have their number 18, i. e.,
$342 \div 19$; and the discharge for one being 289.14 cb. ft.,

the total discharge is 18 × 289.14 = 5204.52 cb. ft. per sec., or 312,271 cb. ft. per minute. Now the back-water containing 15,420,000 cb. ft., it could be discharged in about 50 minutes (15420000 ÷ 312271). It is essential that the back-water should be discharged rapidly before the rising tide diminishes the force of the artificial scouring action.

(x.) The widely different statements as to the efficiency of hydraulic prime movers, some being asserted to give as high as 80, and others 60 per cent. of the power used, may, perhaps, be traced to a false estimate of the actual discharge ; for unless this be gauged, it must be calculated, and some coefficient used. In the case of undershot wheels with sloping sluices, as in Poncelet wheels, the bottom and sides being in continuation of the channel of supply, the coefficient is 0.74 when the sluice is inclined 1 base to 2 height, and 0.80 when 1 to 1 (Claudel, Aide Memoire, p. 78, § 100). If we had taken it 0.62, and with a six feet fall, the sluice being supposed 6 ft. wide and raised 1 ft., we have—

$$0.62 \times 6 \times 8.024 \times \sqrt{6} = 73 \text{ cb. ft. ;}$$

had the modulus been taken as 80 per cent. from this discharge : it would, in reality be but 67 per cent. found by the proportion 0.74 : 0.62 : : 80 : 67.

(xi.) If the modulus of a water-wheel be estimated at 88 per cent. with a coefficient of discharge of 0.65, the wheel being 7 ft. broad, and the sluice, which slopes at 1 to 1, raised 0.75 ft., the head being 5.5 ft. : required the true modulus.

Here $0.65 \times 0.75 \times 7 \times 8.024 \times \sqrt{5.5} = Q = 64.2$ cb. ft., hence (as 0.80 − 0.65 = 0.15),

$$Q (1 + 0.15) : 64.2 : : 88 : 76.56 \text{ per cent.,}$$

the true discharge on the wheel being 73.8, that is,

$$64.2 \times 1.15.$$

(XII.) *Relative Level in two Vessels communicating by a submerged Orifice.*—Let a cistern, A, receive a constant supply of water, and discharge it into a vessel, B, through a, which finally discharges into the air : the orifice at b is 1.0 foot horizontal by

Fig. 34.

0.2 feet, the charge H upon the centre 1.25 ft., and $m = 0.62$; hence $Q = 0.62 \times 8.024 \times 0.2 \times 1.118 = 1.11238$ cb. ft. per sec. which must equal the supply received by A, and transmitted through a to B. Now a is 0.8 ft. by 0.1 ft., the sluice being capable, however, of being raised to 0.5 feet ; and hence the charge upon it, reckoning from the surface of B, is equal to $\left(\dfrac{1.11238}{0.62 \times 8.024 \times 0.08}\right)^2 = 7.806$ ft.; and—as we should expect—the square roots of these charges are inversely as the areas of the orifices ; that is, 2.794 : 1.118 : : 0.2 : 0.08.

Hence, if we suppose the constant supply to be so increased as to raise the surface of the water in A one foot above its level in the last case, we may determine the corresponding rise in B, and also the additional quantity that has been supplied. The total vertical height above the centre of b is now 1.25 + 7.806 + 1 = 10.056, which has to be divided into parts whose square roots have the ratio 0.2 to 0.08, that is, of the areas of the orifices. Now $(0.2)^2 : (0.08)^2$ being as 4.00 to 0.64, we have—

$$4.64 : 0.64 : : 10.056 : \frac{10.056 \times 0.64}{4.64} = 1.387$$

for the surface of B above b, and 10.056 − 1.387 = 8.669 for the surface of A above that of B, and the quantity received in A is now 1.172 cb. ft. per sec. The rise of 1 foot in A corresponded, therefore, to one of (1.387 − 1.25 =) 0.137 feet in B.

If we suppose the surface of A *lowered* 1 foot, then B descends 0.1388 ft., and the constant supply is now 1.05 cubic feet per second. Hence the total range of B is only 0.277 feet for the corresponding change of 2 feet in A.

(XIII.) The time of filling a lock on a navigable canal consists of two distinct intervals: one, the time of filling up to the centre of the sluices; the second, that of raising the surface up to the level of the upper reach. The length of a lock being 115.1 ft., and breadth 30.44 ft., the horizontal area is 3503.6 square feet, and the vertical depth from centre of sluice to lower reach 1.0763 feet, the charge being 6.3945 feet; hence, the cubic content of the lower portion, that is, the value of Q', is 3771 cubic feet; the area of the two sluices 2 × 6.766 sq. feet = 13.532 sq. feet; and the charge on centre, as above, 6.3945 feet; the value of m, assumed by D'Aubuisson, being 0.548. From some experiments on the Canal of Languedoc, it was found that when two sluices were opened in the gates, the discharge was not double that given when only one was used: it was found, in fact, to be about an eighth part less, which reduces m from 0.625 (§ 24) to 0.548. We have therefore—

$$\frac{Q'}{m \cdot S \cdot \sqrt{2g} \cdot \sqrt{H}} = T = \frac{3771 \text{ cb. ft.}}{0.548 \times 13.532 \times 8.024 \times \sqrt{6.3945}} = 25''$$

99. *Determination of the Charge necessary to give a certain Quantity with a given Value of S.*—To determine the head necessary to give a certain discharge, we have but

to solve $Q = mS \sqrt{2g} \sqrt{H}$ for H ; and hence—

$$\left(\frac{Q}{mS \sqrt{2g}}\right)^2 = H.$$

(XIV.) Required the head necessary to give 7.85283 cb. ft. per sec. through an orifice 0.5 feet square, m being equal to 0.625. Here—

$$\left(\frac{7.85283}{1.24175}\right)^2 = 6.324^2 = 40 \text{ feet,}$$

or 2 (log. 7.85283 – log. 1.24175) = log. H ; that is—

2 (0.8950245 – 0.0940167) = 1.6020156 = log. of 40.

If the orifice had been 0.75 feet square, determine the charge necessary to give the same discharge as in the last example, namely, 7.85283 cubic feet per second. Here—

$$\left(\frac{7.85283}{2.821}\right)^2 = \left(2.7837\right)^2,$$

and

$$2 \log. 2.7837 = 2 (0.4446224) = 0.8892448,$$

giving

$$H = 7.748984 \text{ feet} = 7.749 \text{ ft.}$$

What additional head would each orifice require to discharge 10 cubic feet per second, the coefficient remaining 0.625 ?

Here $7.85283 : 10 :: \sqrt{40} : \dfrac{63.24}{7.85283} = $ log. 63.24 – log. 7.8528 (= 1.8009919 – 0.8950245) = 0.9059674 = log. of 8.053, which is the square root of the charge required, whose value is therefore 64.85 ft., and, deducting 40, we have the increase of head equal to 24.85 feet.

And, $7.85283 : 10 :: \sqrt{7.748984} : \dfrac{27.837}{7.85283} = $ log. 27.837

− log. 7.85283 (= 1.4446224 − 0.8950245) = 0.5495979 = log. of 3.5448, which is the square root of the charge sought, 2 × 0.5495979 = 1.0991958 = log. of 12.575, from which deducting 7.748984, we have the additional head in this case equal to 4.826 feet.

(xv.) Calculate the head that is equivalent to the difference between the coefficients 0.600 and 0.950; that is, having the discharge under certain data, with $m = 0.950$; determine what additional head would be required to give the same discharge when $m = 0.600$. Thus, let the charge on the centre be 8.55 feet, the orifice circular and 0.045 feet diameter, and so nearly the form of the *vena contracta* that the coefficient rises to 0.950; we have there-fore—

$$S = (0.045)^2 \times 0.7854 = 0.00159, \text{ also } \sqrt{8.55} = 2.924$$

and $Q = 0.95 \times 0.00159 \times 8.024 \times 2.924 = 0.03544$ cb. ft., and the head necessary to give this discharge with $m = 0.6$ is found (as $mS\sqrt{2g} = 0.6 \times 0.00159 \times 8.024 = 0.007655$) by—

$$\left(\frac{Q}{m \cdot S\sqrt{2g}}\right)^2 = \left(\frac{Q}{0.007655}\right)^2 = H, \text{ or, by logarithms,}$$

$$\left(\frac{0.03544}{0.007655}\right)^2 = (\bar{2}.5494937 - \bar{3}.8839452) \times 2 = 1.331097$$

$$= \text{log. } 21.43 \text{ feet.}$$

Thus 21.43 − 8.55 = 12.88 feet is the additional head or pressure required to discharge the same volume of water through the orifice in a thin plate that was discharged with 8.55 feet pressure through an orifice nearly of the true form. Thus, the accelerating force due to this form, when compared with the thin plate, is measured by a pressure equal to more than one-third of the weight of the atmosphere.

100. *Results of the Suppression of the Contraction on part of the Perimeter,* §§ 25 *to* 26.—A sluice 3 feet square, and with a charge on the centre of 12 feet, has, from the thickness of the frame, the contraction suppressed on all sides when fully open; but when partially opened, the contraction exists on the upper edge, that is, against the bottom of the gate, which is formed of a thin plate of metal. Required the discharge when opened 1 foot, and also 2 feet, and when fully opened.

(XVI.) When opened 1 foot high, the total perimeter is 8 ft., and the part on which the contraction is suppressed is 5 feet : hence—

$$\frac{n}{p} = \frac{5}{8}.$$

Hence, from the formula (§ 26)—

$$m \cdot S \cdot \sqrt{2g} \sqrt{H} \left(1 + 0.152 \frac{n}{p} \right),$$

m being supposed 0.608, we have for the discharge—

$$0.608 \times 3 \times 8.024 \times \sqrt{13} \left(1 + 0.152 \frac{5}{8} \right) = 57.77 \text{ cb. ft.}$$

per second; the two last factors being 3.605 and 1.095.

When opened 2 feet high, the total perimeter is 10 feet, and the contraction suppressed on 7 feet; so that—

$$\frac{n}{p} = \frac{7}{10}$$

and—

$$0.152 \times \frac{7}{10} = 0.106,$$

also $\sqrt{H} = \sqrt{12.5} = 3.5355$. Hence the discharge in this case is $0.608 \times 6 \times 8.024 \times 3.5355 \times 1.106 = 114.45$ cubic feet per second.

K

When the sluice is fully opened, the total perimeter is 1.2 feet and $\frac{n}{p} = 1$; so that the discharge is—

0.608 × 8.024 × 9 × 3.464 × 1.152 = 175.9 cb. ft. per sec.

101. *Questions upon* §§ 44 *to* 52, *formula* $\frac{2}{3} ml H \sqrt{H}$ $\sqrt{2g}$ *and* $\frac{2}{3} ml \sqrt{2g} (H \sqrt{H} - h \sqrt{h}) = Q$.—We derive from Captain Baird Smith's work on " Italian Irrigation " many examples whose solution is given by the formulæ in these sections. It appears in the irrigated districts to be a matter of great importance to determine both a unit of volume of water, and also some means of measuring it so as to regulate the due supply to each proprietor, that, on the one hand, the Government, or individuals having the ownership or management of the canals of irrigation, may not be defrauded ; nor, on the other hand, the proprietors of the land to be irrigated suffer any injustice. Many very different units are found in use at present. In the earlier periods of the construction of these canals, a fixed *area* or orifice, opened in the side of the canal, was alone used, without any reference to the head or charge under which the water issued : to this was subsequently added the condition of a fixed charge as well as a fixed area, but without any mechanical arrangements for insuring this constant pressure, which, from the continual variations in the level of the surface of the water in the canal, was absolutely requisite ; however, unless the excess or deficiency complained of was more than the eighth part of the total volume specified in the grant no complaint was admitted. Several other simple and beautiful contrivances, described further on, are now, however, in use on many works of irrigation, which meet this difficulty, sufficiently, at least, for practical purposes.

Taking these different measures or units in the order of date, as nearly as possible, it appears that on the Canal of Caluso the unit called *ruota*, or wheel, was defined to be the quantity passing through an opening whose area is 1 foot square,—this foot being equal to 1.6702 feet lineal English—the upper edge of the outlet being what is locally termed *a fior di acqua*, or level with the surface of the canal or reservoir, the discharge hence taking place under no pressure. The volume discharged by the *ruota* is estimated by the Piedmontese engineers at 12.05 cubic feet English per second; but obvious sources of error and discrepancy arise from the reference being solely to the superficial area: first, on theoretical considerations; and secondly, on practical. For we may evidently have a number of orifices with very different discharges and figures, all fulfilling the condition of a constant area. The Piedmontese foot being divided into 12 inches, the value of an inch is 1.6702 English inches. The following short Table, limited to whole numbers of inches, will illustrate this point (all in Piedmontese inches) :—

Linear Dimensions applicable to a Ruota.

	Height *H*.		Breadth *l*.		Constant area.	Perimeters.	Ratio of sides.
1.	12	×	12	=	144	48	1 : 1
2.	16	×	9	=	144	50	1 : 0.5625
3.	18	×	8	=	144	52	1 : 0.444'
4.	36	×	4	=	144	80	1 : 0.111'
5.	48	×	3	=	144	102	1 : 0.062
6.	72	×	2	=	144	148	1 : 0.0277

On examining this Table, we see the heights vary from 1 to 6, and the perimeters from 1 to 3, the area remaining constant. Many outlets are instanced, showing that this unit was actually used with a great variety of figure in many grants of water. But, secondly, a

K 2

Examples and

practical objection against any measure whose outlet is level with the supplying surface arises from the fact that this surface is always changing, not only from the variations in the level of the river supplying the canal, but from the hourly changes in the demand for water from the canal by the various land-owners along its length.

(XVII.) Calculate the discharge which would be given by Nos. 1, 3, 5, and 6, of the above Table,—m being taken equal to 0.62, the foot being 1.6702 feet English, we have $\frac{2}{3} \times 0.62 \times 8.024 = 3.315$, and $l \times H = 2.789$ sq. ft. common to each ; and thus—

For No. 1, we have $3.315 \times 2.789 \times \sqrt{1.6702} = 12.00$ cb. ft.

„ No. 3, „ $3.315 \times 2.789 \times \sqrt{2.505} = 14.63$ „

„ No. 5, „ $3.315 \times 2.789 \times \sqrt{6.6808} = 23.90$ „

„ No. 6, „ $3.315 \times 2.789 \times \sqrt{10.02} = 29.26$ „

Or generally, the discharge varying as $l \cdot H \cdot \sqrt{H}$, and $l \cdot H$ being constant, it is evident that it increases as $\sqrt{H}$; so that, by increasing the depth indefinitely at the expense of the width l, we increase the discharge. Thus, let $H = 16$, the log. of $\frac{2}{3}mlH\sqrt{2g}$, that is, of $(3.315 \times 2.789 =)$ 9.2455 being 0.9659304, we must add to it half the log. of H for the log of the discharge : half the log. of 16 is 0.6020600, and adding, we have $1.5679904 = $ log. of 36.982 cb. ft.

102. *Questions on* $\frac{2}{3}\ ml\ \sqrt{2g} \cdot (H\sqrt{H} - h\sqrt{h}) = Q$, *the Italian dimensions being all reduced to English measures.*

(XVIII.) Ignazio Michellotti having determined to modify the mode of measuring a *ruota* introduced by his father, F. D. Michellotti, which had the upper edge level with the surface of the supplying canal, and was estimated to give a discharge equal to 11.83 cb. ft. per sec. defined the *uncia* or inch of water to be that flowing through an orifice 0.5567 feet high, 0.41755 ft. wide, and having a pressure on the upper edge of 0.5567 ft. This he supposed would give the twelfth part of 11.832 cb. ft., or 0.986. Calculate its true value : *m* being 0.600, we have then $H = 0.5567 + 0.5567 = 1.1134$ feet, and $l = 0.41755$ feet. *Ans.* 1.02 cb ft.

(XIX.) The measure used on the Canal Lodi was defined to be 1.12 ft. by 0.12416′ ft. wide, with a charge on the upper edge 0.32 ft., and these dimensions were supposed to give 0.77 cb. ft. per sec. Here $H = 1.12 + 0.32 = 1.44$ ft., and $l = 0.12416'$ ft. *Ans.* 0.6165 cb. ft. per sec.

(XX.) That used on the canal of Cremona was 1.31816′ ft. high by 0.131 ft. wide, having a head also 0.131 ft., and estimated to discharge 0.88 cb. ft. Hence—

$$\tfrac{2}{3}0.6 \times 0.131(1.449\sqrt{1.449} - 0.131\sqrt{0.131})8.024 = 0.715 \text{ cb. ft.}$$

(XXI.) That of Crema was 1.276 ft. high, 0.1275 ft. wide, a charge of 0.255 ft. : calculate the discharge.

Ans. 0.7225 cb. ft. per sec.

(XXII.) The Sardinian Civil Code determines the unit in which all grants of water should be expressed thus :—
" The measure or *modulo* (Fig. 35) is that quantity of water which, under simple pressure, and with a free fall, issues from a rectangular quadrilateral opening, so placed that two of its sides shall be vertical, having a breadth of 0.6562 ft. (English measure), and a height also of 0.6562 ft. It shall be opened in a thin wall (or *plate-parete*), against which the water stands, with its upper

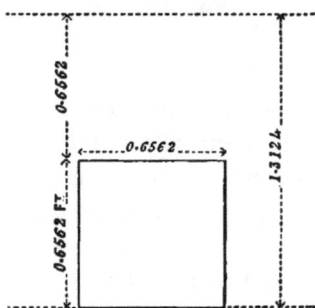
Fig. 35.

surface perfectly free, at a constant height of 1.3124 ft.
(= 2 × 0.6562) above the lower edge of the outlet." It is
required to calculate the value of this unit in cubic feet
per second. We have therefore l = 0.6562, and H and h
being 1.3124 and 0.6562 respectively—

$$\tfrac{2}{3} \times 0.6 \times 0.6562 \ (1.3124 \ \sqrt{1.3124} - 0.6562 \sqrt{0.6562}) \ 8.024$$
$$= 2.046 \ \text{cb. ft. per sec.}$$

When grants are made for more than one *module*, the
only dimension which varies is the breadth of the outlet,
the height and pressure remaining in all cases invariable :
two *modules* would have a breadth of outlet of 1.3124 ft.,
three would have 1.9686 ft., and so on.

103. *Description of a Piedmontese Outlet* ("Italian Irrigation," pp. 21, 22, vol. ii.).—" AB (Fig. 36) is a portion
of the transverse section of the supplying canal ; the
first part of the measuring apparatus is the sluice, which

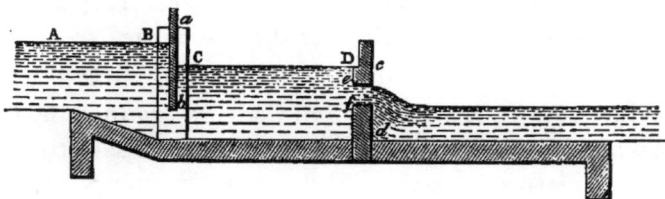

Fig. 36.

consists of masonry side-walls, and a gate of wood, working vertically. The dimensions of this primary outlet
are not rigidly fixed, its object being merely to admit a
larger or smaller supply into the chamber CD. The
sluice is established in the bank of the canal, at such
point as may be fixed upon by the canal authorities, or
most convenient for the land-owner. Its sill is sometimes
on the same level as the canal bed, sometimes above it,
and very frequently as represented in the diagram.

There is a fall in front of the outlet, so as to draw the water towards it. For a length of from 40 to 50 feet from the sluice, the bed of the channel is made perfectly horizontal, paved with masonry or cut stone, the upper surface of which is on the same level as the sill of the sluice. At a distance from the outlet, ranging from 16 to 32 feet, is fixed the partition or slab of stone *cd* in which the regulating or measuring outlet *ef* is cut, the height of which is fixed at 0.56 ft., while the breadth varies with the number of units or inches to þe given, each inch being represented by 0.42 ft. of breadth. The lower edge of the measuring outlet is ordinarily placed at 0.819 feet above the level of the flooring of the chamber CD. A small return cut in the inner face of the slab, at a height of 0.28 ft. above the upper edge of the outlet, indicates the constant level of the water necessary to insure the established pressure. This height is maintained by the raising or lowering, as may be requisite, of the sluice at the entrance of the chamber.

(XXIII.) Calculate the value of a grant of three inches of water from this structure. We have $H = 0.56 + 0.28 = 0.84$; hence—

$$3 \times \tfrac{2}{3} \times 0.6 \times 0.42 \times (0.84 \sqrt{0.84} - 0.28 \sqrt{0.28})\ 8.024$$
$$= 2.514 \text{ cb. ft per sec.}$$

104. *Description of the Modulo Magistrale of Milan.*— This module, as applied upon the Naviglio Grande, which in a course of 31 miles from its head on the River Ticino to the city of Milan, distributes 1851 cb. ft. per second, is in its principle identical with that already described (§ 103). For the interesting history of this canal, and the gradual improvements in the management of the grants of water, we refer to "Italian Irrigation," vol. i., pp. 203, 228; vol. ii., pp. 36, 56. The honour of the discovery is due to Soldati, of Milan, about the year 1571,

who invented it in answer to an invitation from the magistracy of that city to architects and engineers to design a measuring apparatus.

The unit fixed upon, called the *oncia magistrale*, had the following dimensions (Fig. 37):—Height, 0.655 ft.; breadth, 0.3426′ ft.; with a constant pressure of 0.32944

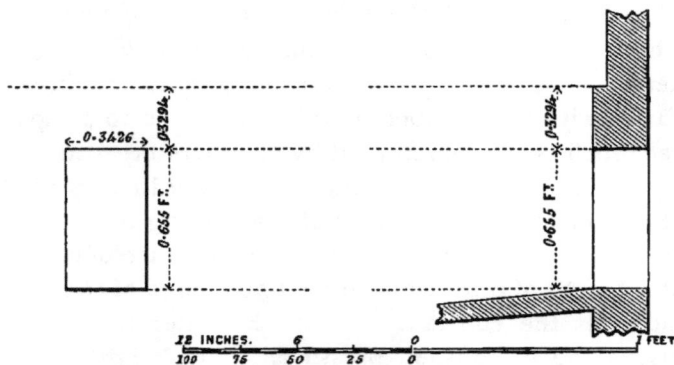

Fig. 37.

ft. above the upper edge of the outlet. When one outlet is designed for the discharge of several water-inches, the breadth only varies, in the proportion of 0.3426′ ft. for each additional water-inch, the height and pressure remaining constant, as in Fig 38, which shows an outlet for six water-inches. The outlet is cut with care in a

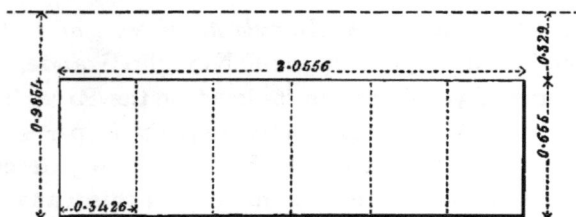

Fig. 38.

single slab of stone. To preserve it from being tampered with, an iron rim is fixed upon it, of the exact dimensions corresponding to the discharge. They ought invariably to be cut in a simple plate, with no arrangement of any kind to increase the volume beyond that due to pressure alone. The thickness of the slab varies somewhat with the dimensions of the outlet; but in a rigidly exact *module* this dimension should be fixed as well as all the others. These are the conditions applicable to the measuring outlets, the discharge from which is—

$$\tfrac{2}{3} \times 0.6 \times 0.34266' \, (0.98444 \, \sqrt{0.98444}$$
$$- \, 0.32944 \, \sqrt{0.32944}) \, 8.024 = 0.866 \text{ cb. ft.}$$

To illustrate the other arrangements of the *modulo*, the horizontal and vertical sections (Figs. 39 and 40) are given from the same work.

PLAN.

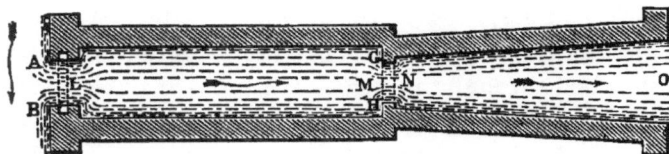

Fig. 39.

The sluice AB (Fig. 39) is placed on the bank of the canal of supply, with the sill CD (Fig. 40) on the same

SECTION.

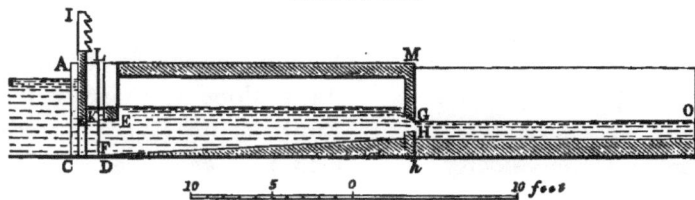

Fig. 40.

level as the bottom of this canal. It is formed of two

side-walls or cheeks, of good masonry, in brick or stone, with a flooring generally of the latter material. To prevent erosive action, the bed of the canal, for such distance as the force of the current may render necessary, is paved with slabs of stone or boulders, both above and below the head. The sluice gate is usually made of the same breadth as that of the measuring orifice GH (Fig. 39), while its height is regulated by that of the sluice itself. The sluice-gate or *paratoja* IK (Fig. 40) works in grooves, and is fitted with a rack and lever, by which it can be readily raised or depressed at pleasure. As the surface level of the canals of the Milanese varies comparatively little, the upright of the sluice has a small catch in iron or wood attached to it, by which it is kept at a fixed height, corresponding to the requisite pressure on the original orifice GH (Fig. 40). This little catch is locally termed the *gatello*; and as it is provided with a lock and key, the latter of which is intrusted to the guardian of the canal, the proprietor of the water-course supplied through the *module* is supposed to be restricted to his legitimate supply, and probably is so within reasonable limits, provided always that the guardian is incorruptible. In the rear of the sluice-gate, at the head, is placed the first chamber LM (Figs. 39 and 40), called the *tromba coperta*, or covered chamber. Its length is equal to very nearly 20 feet, with a breadth varying according to the size of the head-sluice, which it exceeds by the fixed quantity of 0.82 ft. on each side, or 1.64 on the entire breadth. The bottom of the covered chamber DH (Fig. 40) is formed with a slope to the rere, the height Hh being 1.3125 ft. English: its object is to diminish the velocity with which the water reaches the measuring outlet GH. Further to assist in effecting this object, the perfect *modulo* is provided with a horizontal top of stone slabs or planks, called the *ciclo morte*,

the under surface of which is at precisely the same
height as the water ought to have over the outlet GH, so
as to secure the fixed discharge, that is, 0.32944 ft. above
the upper edge of GH. It is found that this does reduce
the irregular motion of the water, and so tends to secure
the great object of the *modulo*, that the discharge should
take place under simple pressure, and without antecedent
velocity. To admit of ready inspection of the height of
the water within the covered chamber, the following arrangements are made :—The entrance to the chamber is
covered with a stone slab of convenient thickness, shown
in section at E (Fig. 40), the lower surface of which is
precisely on the same level as the upper edge of the outlet GH. The height of the slope H*h* being 1.3125 ft.,
and that of the outlet GH being 0.655, the surface of the
slab at E should be 1.9675′ ft. above the sill of the head
CD. An open groove LD is made in the masonry, large
enough to admit a graduated rod or measure ; and when
the water stands at a height of—

$$(1.9675' + 0.3234 =) \ 2.297 \text{ ft.}$$

above the sill at D, it is known that the proper head of
pressure exists at GH. As it is found to be greater or
less, the sluice is depressed or raised, so as to adjust the
pressure to the fixed standard. The slab of stone in
which the measuring outlet is cut being fixed at GH
(Figs. 39 and 40), immediately in rere of it there is placed
the *tromba scoperta*, or open chamber. Its breadth at N
(Fig. 39) is two local inches (0.3275 ft. English), greater
on each side than that of the measuring outlet, or on
both sides 0.6550 ft. Its total length NO is very nearly
17.75 ft. English. Its side-walls, which are perpendicular, like those of the covered chamber, have a splay outwards, so that the breadth at O is 0.9825 ft. greater than
at N, or 1.31 ft. in excess of that of the regulating outlet

GH, being the same as that of the covered chamber
throughout. To insure the free run of the water from
GH, the flooring of the open chamber has a drop or fall
of 0.1633 ft. at H, and an equal quantity distributed uniformly between H and O (Fig. 40). There is therefore
a total fall from the under edge of the measuring outlet
to the end of the open chamber of 0.3275 ft. or, as the
length is 17.72 ft., of 1 in 54. When the water reaches
O, it enters the channel of distribution for the use of the
consumers : generally the point O, and the bed of the
channel, which is carried on at the usual inclination, are
upon the same level, but sometimes there is a fall.

105. From the preceding details, it appears that the
modulo magistrale has a total length of nearly 37.75 ft.
English, and a breadth variable according to the quantity of water it is intended to measure. If a single water-
inch, for instance, be granted, the breadth of the covered
chamber would be 2.12835 feet, and that of the open
chamber 1.145835 feet at its upper, and 2.12835 at its
lower extremity. The flooring of the former rises 1.31
ft. to the rere, while that of the latter falls 0.3275 ft. in
the same direction. It is essential to the effective operation of the regulating sluice in the *modulo magistrale*
that there should be a difference of level between the
water in the canal and in the apparatus of at least 0.655
ft.; and as the height of water in the latter must be 2.297
ft., the depth of water in the canal of supply must necessarily be not less than the sum of these numbers, or
2.952 ft., very nearly 3 ft.

It is curious to reflect that this apparatus was invented
empirically by Soldati, in 1571, so many years before
the discovery of the Toricellian theorem, which must be
placed in the year 1643, when that philosopher showed
that the laws of running water were identical with
those of falling bodies, the foundation of all our know-

ledge of Hydraulics. This is not the only instance in which the practical sagacity of the engineer has anticipated the discoveries of theory.

Two essential objects are supposed to be fulfilled by these arrangements :—1st. To maintain on the measuring outlet a constant pressure ; and 2nd. To make this pressure as much as possible the sole force influencing the discharge, that is, that the water have no velocity antecedently. The first is secured by the mechanical arrangements at the head,—the sluice with its rack, lever, &c., and to a certain extent the *cielo morte*. By raising or lowering this sluice the level of the water in the covered chamber is maintained, independent of the variations in the surface of the external canal. The second by the interior arrangements,—the covered chamber with its fixed top, and floor sloping up to the outlet ; while the free passage of the water is secured by the open chamber, with its small fall at the head and continued inclination at the bottom.

106. The differences in the estimates of the quantity of water discharged by the *modulo magistrale*, as given by different Italian engineers, are very remarkable, considering the great attention that has been paid to the theory and practice of Hydraulics in that country. De Regi gives it as 1.42 cb. ft. per sec.; Breschetti states the average result of experiments on the Muzza Canal to give 1.57 cb. ft. per sec.; Mazzeri estimates it as low as 1.21 cb. ft. ; Brunacci at 1.46 ; while the Department of Public Works in Lombardy considers it equal to 1.64 cb. ft. per sec. The extremes, we see, are 1.21 and 1.64 cb. ft. per sec.,—a difference of 0.43 cb. ft., between a third and fourth of the total discharge. Captain Smith accounts for this great difference by stating—"That the estimate of the Government is founded on the experience of the results on the great canals, where the outlets are almost uniformly of

large dimensions." (pp. 222, 223, vol. i.) Now it is certain that, all other circumstances being alike, the quantities of water discharged from large are proportionally
greater than those discharged from small outlets. Hence
the *oncia magistrale,* as determined by experiments with
the former, has a decidedly higher value than when determined by the latter.

The cause of this is clear. To give a discharge of,
say, six water-inches, the breadth of the outlet is made
six times that for one inch, the height and the pressure
remaining in both cases the same. The proportion between the sectional area and perimeter of the outlets
becomes, however, materially altered, and the influence
of the perimeter in effecting the contraction of the vein
diminishes gradually as the size of the outlet increases;
and in a similar proportion the discharge becomes greater.
To elucidate this, it may be remarked, that in an outlet
for one *oncia magistrale* the ratio of the section to the
perimeter is as 1 to 23.33; for two, as 1 to 16.66; for four,
as 1 to 13.33; for eight, as 1 to 11.66; for ten, as 1 to 11.33,
or about half what is for one *oncia ;* for twenty *oncia,* as
1 to 10.66, and so on; and there are real differences of
discharge due to the variable ratios now given.

Very serious pecuniary loss may consequently be the
result to the proprietors of the canal or the consumers of
the water. It appears (vol. i., pp. 226, 227) that for
summer irrigation each cubic foot per second is capable
of irrigating 61.8 acres, and that the annual rent of this
quantity, summer and winter, is £13 5s.; the difference
of 0.43 cb. ft. between the highest and lowest estimate of
the discharge of the *modulo magistrale* is worth £5 13s.,
and would irrigate 26 acres at the above rate.

The recognition of the differences between the discharges of large and small outlets was very early made
in Lombardy. In the *module* of Cremona, invented in

1561, no single outlet was allowed to exceed 1.31 ft. high
by 3.18 ft. broad, equal to about 12 or 13 water-inches.
In the Milanese single outlets have been restricted for
nearly three centuries and a half to discharges of from 9
to 12 *once*. In Piedmont they have been more careful,
and have there limited single outlets to 6 *once*, which,
by general consent, seems to be the most approved size
for diminishing to the utmost the error due to the in-
equality of discharges from large and small openings.
For practical purposes, therefore, and taking the mean
of the various estimates of the value of the *oncia magistrale* just adverted to, it may be considered as equal to
very nearly 1½ cb. ft. per sec.

107. Another mode of insuring a constant discharge
through an orifice having a charge subject to variation
has been brought into use by the late Mr. Thom,- an hydraulic engineer of great eminence. It attains this

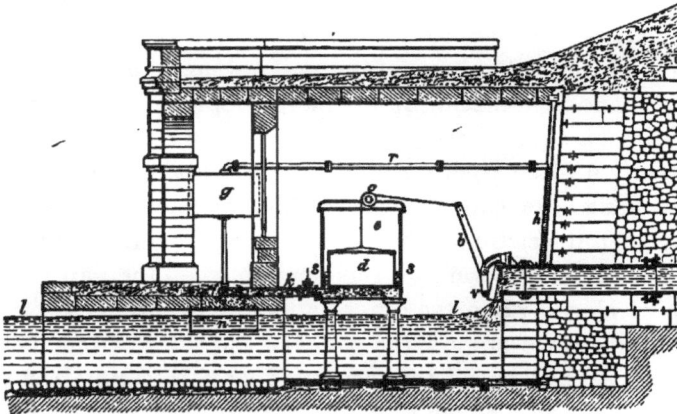

Fig. 41.

object by mechanical means chiefly. Fig. 41 represents a vertical section of the regulator at the Gorbals
Waterworks, near Glasgow. The discharge pipe from

the reservoir is on the right-hand side. If the quantity drawn off by the town or mill to be supplied should increase, then the level of the surface *l, l* will descend ; and the apparatus must be such that it may permit a larger quantity to pass through the pipe, and *vice versâ.* Again, if the level of *l, l* should remain constant, and, from an increased or diminished rainfall, that of the reservoir rise or fall, then this apparatus should be so constructed as to adjust the orifice of the discharging main pipe that it deliver only that constant quantity carried off from the receiving basin, and needed for the town or mill-works.

Fig. 41 gives a longitudinal section of the detail of the regulator : *d* is a moveable cast-iron cylinder or float attached at top to a chain passing over the pulley or wheel *c*, and surrounded by a fixed cylinder of a diameter slightly larger, containing water, and represented in section at *e*. The other end of this chain is fixed to the bent lever *b*, working freely on a stud carried by two cast iron brackets screwed to the extremity of the pipe passing through the base of the embankment of the reservoir, and terminating in a square mouth-piece, faced to receive a square hinged flap-valve, *a*, which is retained in any desired position by the lower and shorter arm of the bent lever which works against the back of the valve by an anti-friction roller at *v* ; the inner cylinder *d* must be loaded with weights sufficient to keep the flap-valve quite closed when the outer cylinder *e* is empty.

Now if we suppose the water in the outer cylinder *e* to stand at the level *s s*, the cast-iron float being immersed to a certain depth below this surface, part only of its weight, acting by the chain upon the bent lever *b*, will press against the square flap-valve and thus partially open the mouth of the main-pipe, restricting the discharge through it to the desired quantity. Suppose, then, that from any circumstances this discharge should become too small, and therefore the surface *l, l descend*, it will then be necessary that the self-acting apparatus should be such as to permit the valve to open, and therefore, also, the cast-iron float to *rise*, which it will do if the water-level in the outer cylinder be made to rise; for then the cast-iron float becomes specifically lighter, and presses with a less force upon the valve *a*, which immediately yields to the pressure of the water issuing through the discharge-pipe, and thus permits a greater quantity to escape.

If, on the other hand, the quantity discharged had been too great, and thus the surface *l, l rise,* it will be necessary that the cast-iron float *descend*, and thus press the flap-valve closer upon the square face of the discharge-pipe. This it will do if the water in the outer cylinder be made to fall; for thus the float becomes specifically heavier, and sinks, closing the flap-valve *a*: so that we have to devise such mechanical arrangements that when the discharge is too small, the water surface in the cylinder *e* shall rise, and when too great that it shall descend.

This is effected in the following manner:—A small closed cistern, *g*, is placed at the side of the portico of the entrance door of the building; this is supplied with water by a horizontal pipe, *r*, in communication with the vertical pipe, *h*, placed on the discharging main for the escape of air, which would otherwise collect within it, and greatly impede the discharge.

In all cases of discharge of water through pipes, care must be taken that the air which may collect be readily let off.—*Vide* Buck's Account of the Montgomeryshire Canal Lock; Simms on Public Works in England, p. 8.

The pipe, *h*, must necessarily be carried up the slope of the embankment, and communicate with the air above the level of the highest water in the reservoir. The cistern, *g*, is thus kept constantly supplied with water, and a communication is formed by the pipe *k* between it and the cylinder *e*. In the vertical part of this pipe are fixed two double-beat valves—described below—whose common spindle is fixed to the float *n*, placed in the receiving basin *l*, *l*; now if the surface of the water upon which *n* rests should rise beyond the proper level, then this float, *n*, also rises, and, forcing up the spindle, closes up the upper or discharge valve from the cistern, *g*, and, as the valves are fixed on one spindle, of course simultaneously opens the lower one, so that the water which buoys up the float *d*, in the cylinder *e*, begins to flow out, and the consequent depression of the surface *s*, *s*, causing *d* to descend, partially closes the flap-valve, *a*; and therefore the surface *l*, *l* begins to descend, and with it the float *n*, which necessarily opens the valve which had shut off the water from the cistern *g*, and it, again receiving a supply, *d*, rises, and consequently the flap-valve opens, and thus very soon arrives at a position giving nearly perfect equality between the supply and consumption of water.

In cases when the pressure upon a sluice is not great, the float *n* may be directly connected with the lever which works the sluice.   Fig. 42 represents this simple apparatus : *a*, *a* is the transverse section of the conduit, in which the sluice *b* moves vertically, and is connected by an adjustable link with an oscillating beam *c*, jointed to the top of the short pillar *d*.   The other extremity of this beam

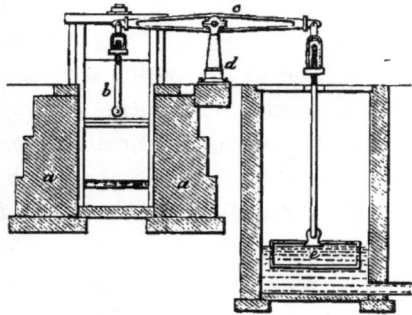

Fig. 42.

is similarly connected to a hollow wrought-iron float

c, which is acted upon by the water whose surface is intended to be preserved at the same constant level, and the supply of which is derived from the conduit *a* ; if then the surface at *e* rise, the sluice is depressed, and the discharge by the conduit lessened, and *vice versâ.* This arrangement is evidently only suited to an open

conduit, in which no great pressure can be brought upon the sluice; if applied to the mouth of a closed pipe with a great head of water pressing on it, the friction in the grooves of the sluice-frame would be so great as to require an enormous float *e*, and the action could not fail to be of an irregular character.

The double-beat valve, invented by Hornblower (Pole on the Cornish Engine, pp. 85–88), is represented fully opened in transverse section, at D, C, Fig. 43, and shut in Fig. 44. It is intended that the water or steam should pass from A to B when the valve is opened, and that the communication between them be intercepted when it is shut. The dark lines at

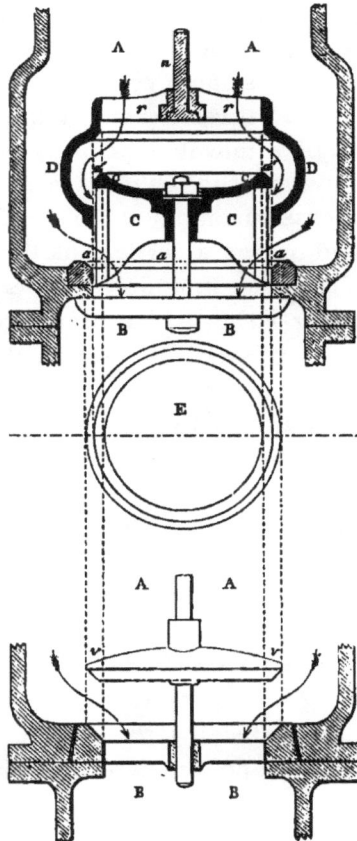

Fig. 43.

D, D represent the movable parts of the valve ; those at C, C indicate the parts that are fixed. The value of its peculiar construction may be best appreciated by

considering the tests of a good valve, which should, in
the first place, evidently afford a large passage to the
steam or water, with a small displacement; and, secondly,
should be capable of being opened with a small force.
These conditions are fulfilled in the double-beat valve,
which consists of a fixed part or seat C, formed by five
partitions, which radiate from a central axis, and are
joined below to a ring, *a*, Fig. 43,
and closed on top by a circular disc, in one piece with the
partitions, and covering the
spaces between them, and also by a movable part, D, the
valve proper, which is a sort
of case surrounding the seat C,
and having a vertical motion,
sliding up and down the exterior edges of the partitions
in C; this case is open on the
top, and connected with its
actuating rod *n*, by the arms *r*, *r*.

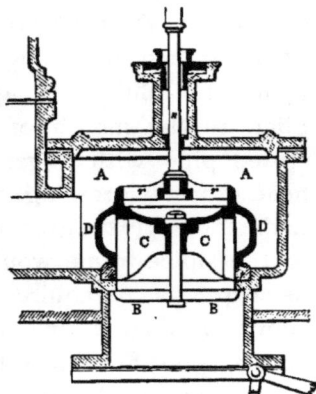

Fig. 44.

When it is at the lowest point of its stroke, and shut,
it bears upon the bevilled or conical surfaces *a* and *a'*,
which have but a very small breadth; when, on the contrary it is raised, as in Fig. 43, it permits the passage of
the water through the different openings shown by the
bent arrows. It is evident that by this arrangement it
is not necessary to raise the valve through any great
height in order to afford a large passage to the water,
thus satisfying the first test mentioned above; on the
other hand, the valve D, being pierced on its upper part
by a circular opening nearly as great as that on the
lower part, the force required to raise it is the excess of
the pressure of the water or steam per square inch in A
over that in B, multiplied into the difference of the

circular areas above mentioned, this difference being evidently the annulus formed by the sum of the horizontal projections of the upper and lower conical surfaces, *a* and *a'*, shown at E in Fig. 43, projected down from the transverse section.

If this valve or case D should have been a simple disc with bevilled edges, as in the lower part of Fig. 43, we should have required to lift or start it a force equal to the excess of pressure in A over that in B, multiplied into the whole circular area of the top of the disc *v, v*; and this would not only have to be provided by the prime mover, but a very greatly increased size and strength given to the rods, joints, &c., which actuate the valve. In a large disc-valve, as, suppose, 12 inches diameter, the area being 113.1 sq. inches, and with an excess of pressure in A above that in B of 15 lbs. per square inch, it would require a force of 1696.5 lbs. to lift it. If in an equal double-beat valve each annulus was $\frac{3}{4}$ inch broad in the horizontal projection, the sum of their areas would be $(12^2 - 9^2) \times 0.7854 = 49.48$ sq. inches ; thus the force required for the starting of such a double-beat valve is less than half that necessary for an equal disc-valve, being $49.48 \times 15 = 742$ lbs., or 954 lbs. in favour of the double-beat valve, and so in proportion for pressures other than 15 lbs.

In the particular case of the valves raised by the float *n*, Fig. 41, it may be, moreover, remarked that the force necessary to raise them has to be applied but for a very short time, the instant it is raised, the pressure on each side is brought to a state nearly that of equilibrium ; the less, then, the resistance to the float at the moment of raising the valve is, the more sensitive it becomes to any alteration in the surface of the water in *l, l*, with an absence of any irregular or jerking motion. The flap-valve *a* is consequently retained more steadily at its proper adjustment.

The woodcut, Fig. 45, illustrates a somewhat simpler arrangement to effect the same object, as in Fig. 41, which has been adopted by Mr. Gale at the Kilmarnock Water Works. It represents a vertical section through the centre line of the valve house, showing half the roof and the end walls, the entrance door being at P, and, at the opposite end, the pipe A from the reservoir enters ; N and

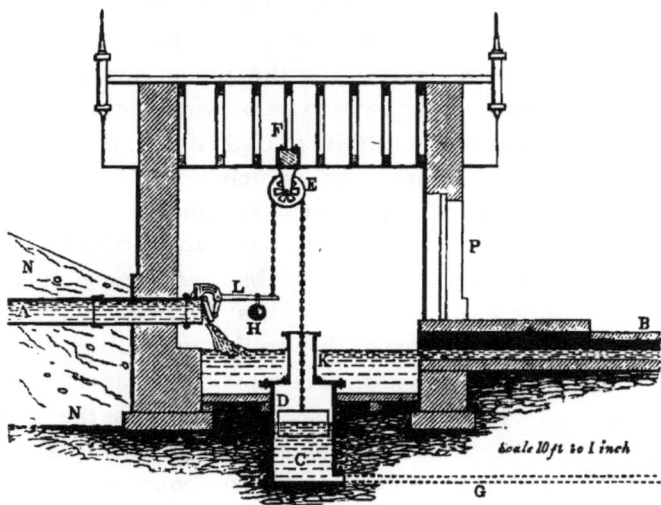

Fig. 45.

N being the section of the foot of the external slope of the embankment impounding the water, and B the culvert conveying it to the filter beds ; the plane of the section is taken transversely to the embankment and perpendicular to its length. The lever L, its support, and the flap-valve are of the same construction as those described above. The weight H is sufficient by its leverage to close the flap-valve and prevent any discharge taking place : all the supply, therefore, must be given by a reduction of the pressure so produced.

A chain attached to the outer end of the lever L

passes over the pulley E suspended at F from a transverse beam shown in section. The other end of the chain is fixed to a float D, working in a cast-iron circular well C; on the cover of it is bolted a tube K, which rises above the highest water level in the valve house. The bottom of the well is in communication with the distributing reservoir by the horizontal pipe G, so that the water stands at the same level in both : if, therefore, the consumption in the town were such as to cause that surface to descend, the float D, becoming less supported, pulls upon the chain, and, lifting the end of the lever L, increases the discharge through A : this increased volume passing down the culvert B to the filter beds soon arrives at the distributing reservoir, and tends to restore its level; on the other hand, if the surface were to rise, the float D, becoming more immersed, loses a portion of its weight, and, therefore, the valve at the end of the pipe A is proportionably closed, and the discharge lessened in correspondence with the lessened consumption. The depth of water in the reservoirs at these works was about 18 feet; at the Gorbals reservoir, about 50 feet, it would not have been possible with this greater pressure to have adopted the simpler arrangement just described; the dimensions of the float D and and the weight H would have been inconveniently increased.

The moderateur lamp affords a most ingenious example, though on a very small scale, of a constant flow of the oil, though the " head," or pressure, varies widely. The annular wick, or Argand burner, is placed on the upper part of the lamp, and is fed with oil from a cylinder which is placed at the lower part of it, and closed at the bottom. The oil is raised from this by the descent of a piston, forced down by the uncoiling of a spiral spring, which is compressed in winding up the piston from the bottom of its course after the former time of use;

an ascending pipe, passing through this, conveys the oil
up to the wick. Now, not only is the spring weaker as
it expands with the descent of the piston, but the vertical height it has to raise the oil also increases : thus, if
were not for the contrivance about to be described, we
should have the brilliancy of the light continually lessen-
ing as the rate of supply of oil to the wick diminished.
This difficulty the celebrated James Watt did 'not quite
surmount when he turned his attention to this subject
(*Vide* Life, pp. 462–465).

A straight wire, or rod, is placed concentrically within
the ascending pipe, of a diameter but little less than it
at its upper and thickest part, and long enough to enter
the movable pipe when at its lowest position ; the lower
part is only a support. The oil in rising is compelled
to pass through the narrow annular space between the
interior of the moveable pipe attached to the piston and
the rod (or *moderateur*) ; from this it results that it meets
with a resistance which causes its upward movement to
be very slow. Now as the moveable pipe and piston
descend, the same length of the *moderateur* is not always
engaged in the pipe: at first, when the spring is strongest,
and the height that the oil has to rise is least, then also
the annular passage is longest, and the resistance to
the ascent of the oil greatest ; and again, when the
piston has descended, and consequently the spring is
weaker, and the height the oil has to be raised greater,
so also the length of this annular space is less, and the
resistance to the ascent of the oil diminished in proportion as the ascending force itself is diminished. By a
tentative process in each particular case—filing a portion
into a flat surface—the needle is adjusted so as to give
a uniform supply of oil, and make the lamp burn with
equable light as long as the spring acts. This principle
is evidently applicable to the discharge of water by
simple modifications.

The "Module" adopted on the canal of Isabella II. is shown in Figs. 46, 47, 48, 49, taken from the work of Lieut. Scott Moncrieff. It consists of a float, M, and a plug, N, suspended from it, which works in a circular orifice in a plate set at the level of the bottom of the channel.

All being contained in a rectangular well of masonry 3.28 ft. by 3.94 and 4.16 feet deep, communicating with the main channel by a lateral opening having an iron grating in front, and covered by a locked iron trap-door, to prevent all tampering with it. The float M, which is formed of brass plate, is shown in plan in Fig. 47, and in elevation, with plug attached, in the upper woodcut, Fig. 46. The surface of the water, and, therefore, the apparatus which rises and falls with it, being supposed at its highest level, and one meter in depth. It is also shown in Fig. 48, at a larger scale, in a vertical central section through AB in Fig. 47, and two of the three central supports are given, carrying a central disc, through which passes the screwed end of the rod that the plug is suspended from; a butterfly nut on the top enables the whole to be adjusted. The plug and the plate, in which is the orifice for the outlet of the water, are of bronze to avoid rust.

ELEVATION.

Fig. 46.

PLAN.

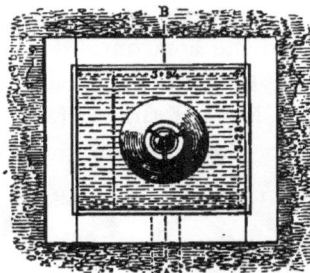

Fig. 47.

The water entering laterally from the canal passes down through the annular space between the plug and plate. From the form of the plug it is evident that this space increases as the level of the water is lowered, and if the area of the annular opening be inversely proportional to the square root of the "charge" or depth of the canal above the bronze plate, then we should have a constant discharge under all the variations in the level of the water flowing down the canal. In Fig. 49 we have represented four horizontal sections taken at the corresponding numbers on the right-hand side in Fig. 48; if the water surface were to descend until the horizontal dotted line at 2 reached the level of the orifice,

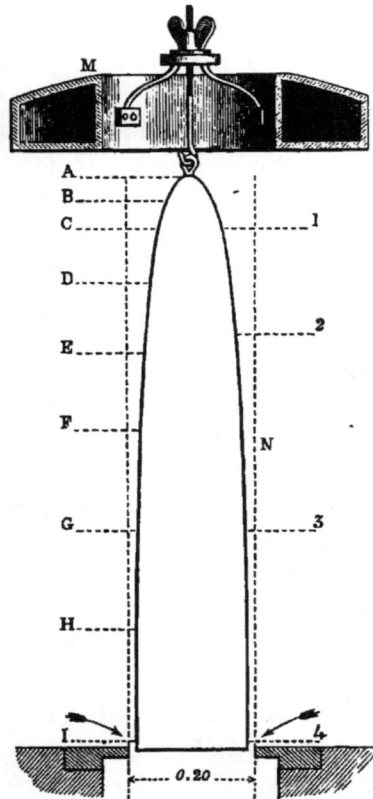

Fig. 48.

then the area for the discharge would be the annular

Fig. 49.

space shown at the corresponding number in Fig. 49;

as the outer circle, representing the circular orifice, is constant, the area for discharge evidently increases as the plug descends.

The objection to this module is, that it involves a considerable loss of head, the level of the water in the culvert, flowing off to the irrigation channels, must be lower than that in the canal of supply by, at least, double the height of the plug, so that it would be inapplicable to the great plains of India, where every inch of level has to be economized. The diameter of the orifice in Fig. 48, which shows the float and plug at a larger scale, is 0.20 metre (= 7.874 ins.), and that of the plug at the same level, when in its highest position, is 0.1653 metre (= 6.51 ins.), so that the area of the annular space in this case is 15.436 sq. ins. ; and, as the coefficient of contraction, from experiment, was found to be 0.63, we have 0.63 × 15.436 = 9.72 square inches = m S, or 0.0675 square feet, which is the opening at 4 in Fig. 49, and the depth of the water over it, or the charge, is 1 metre (= 3.281 ft.). Had the water surface lowered until G, at 3 on the plug, coincided with the orifice at I, then the annular opening is that shown at 3, Fig. 49; at this point the diameter of the plug is 0.1554 ft.

Let D represent the diameter of the orifice in the bronze plate fixed in the bottom of the chamber,

And H the depth of water over it when the main channel is running full.

Let d_4, represent the diameter of the plug at the base, and

d_3, d_2, &c., the respective diameters at the several points so numbered, Fig. 48, between the base and vertex of the plug, when

h_4, h_3, &c., are the corresponding depths estimated from the lowered surface of the channel, these two quantities being so related to each other that the increased area of the annular space may compensate for the diminished charge and give a constant discharge.

From experiments it was found that the coefficient of contraction in this case was 0.63. Let Q represent the unaltered quantity which it is desired to discharge at every different level of the water; then to compute d_4 the diameter of the plug at the base when the charge is H, we have—

$$Q = m \times \frac{\pi}{4} \times (D^2 - d_1^2)\, 8.024 \sqrt{H};$$

the constant multipliers amount to 3.97,

and thus $\dfrac{Q}{3.97} = (D^2 - d_1^2) \sqrt{H}$,

and $\dfrac{Q}{3.97} = (D^2 - d_1^2) \sqrt{h_1}$.

Hence, if we assign successive values to either h_1 or d_1, we obtain the value of the other. If h_1 be given, we have—

$$d_1 = \sqrt{D^2 - \frac{Q}{3.97 \sqrt{h_1}}},$$

and if d_1 be given, we have—

$$h_1 = \left(\frac{Q}{3.97\,(D^2 - d_1^2)} \right)^2$$

The following Table gives the dimensions, in metrès, of one of these modules from Lieut. Moncrieff's work.

It may here be stated that an identical regulator for the flow of gas has long been in use in this country.

The letters refer to Fig. 48 :—

	Diameter.	Depth from Surface of Canal.
At A	0.0000	0.10
B	0.0585	0.12
C	0.0912	0.16
D	0.1211	0.25
E	0.1374	0.36
F	0.1430	0.49
G	0.1554	0.64
H	0.1610	0.81
I	0.1653	1.00 meter
The fixed opening 0.2000		

The self-acting module, of which the details are

ELEVATION.

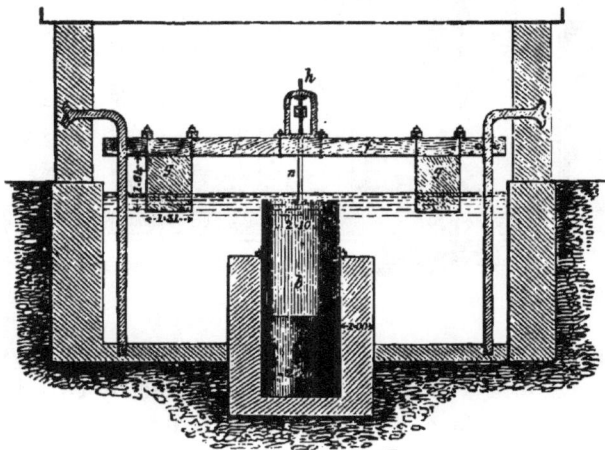

Fig. 50.

given in sectional elevation, Fig. 50 and in plan, Fig. 51, is that used on the Marseilles Canal, and differs en-

tirely in principle from the one proposed for the Isabella
II. Canal, described above, p. 153, in which it is arranged
that as the head of water increases the outlet is diminished,
and *vice versâ.* In the module now considered the outlet remains always the same, and at the same distance
below the surface of the water, with which it rises or falls,
being attached to floats. It will be seen by the engravings, that at the bottom of a masonry cistern connected with the canal,—the bent arrows show the entry
of the water,—there is a circular orifice, *c*, Fig. 51, into
which is accurately fitted, by a water-tight collar, an iron

PLAN.

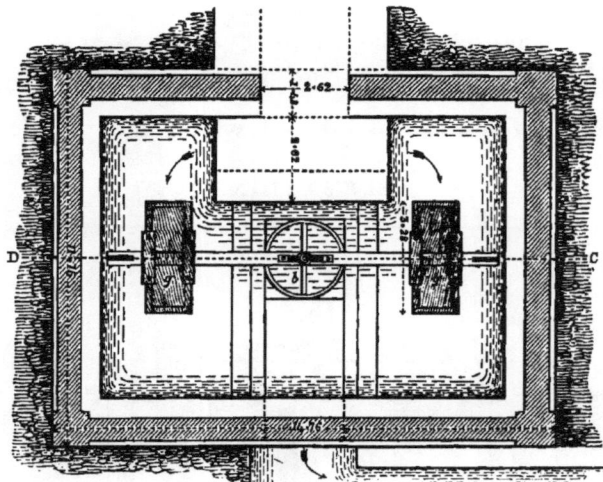

Fig. 51.

cylinder, *b*, open at each end. This cylinder hangs by
the rod *n* to a wooden bar, *f, f,* supported by two floats,
g, g, on the surface of the water, and slips freely up and
down in its collar. By means of a screw at *h* the distance of the upper edge of the cylinder from the bar, and,
consequently, from the water's surface, is fixed, and that

being done so as to give the required discharge, it is never altered. The module is contained in a small locked house. The water below the orifice, after having passed over the circular edge, goes direct to the irrigation channel, as shown by the external bent arrow. It does not require the same loss of level as that of Isabella II.; but it is difficult to believe that the iron cylinder, however nicely adjusted, can always work true, and without either friction or considerable leakage. The silt is observed to collect about this module in considerable quantities. It is, however, stated to satisfy both the canal engineers and the irrigators who pay for the water, so that its working can hardly be very inaccurate, and thus a great object has been obtained. The chief dimensions have been given in feet and decimals.

The water module designed for the Henares Canal in Spain by the eminent hydraulic engineer, J. F. Bateman, is thus described by Lieut. C. C. S. Moncrieff in the work already quoted :—About 4¾ miles from the weir at the head of the canal is the first module for an irrigating channel, branching off from the main canal. Each of these channels is to have a discharge not exceeding 6.22 cb. ft. per second (= 176 litres, nearly 10 cb. metres per minute). The modules are of a rather expensive construction, costing £60 each, and will doubtless do what is intended very efficiently. The self-acting principle has not been tried at all; but the regulation is effected, as shown in Figs. 52 to 56, by means of a very neatly fitting cast-iron sluice, raised and depressed by a screw, and admitting the water into a masonry chamber, out of which it escapes over a bevilled iron edge.

The guard in charge has orders to keep the level of the water to a certain height, denoted by a gauge, in this chamber. To effect this he opens or shuts the sluice according to the fall or rise in the canal. The water passes the sluice with of course a good deal of boiling action, which is completely stopped by a species of masonry grating, M, Figs. 54 and 55, built across the chamber, dividing it into eight passages, each 5.4 inches wide. On the lower side of this partition the water is perfectly still, and drops gently over the iron edge. The opening of the sluice is 1.97 ft. × 1.97 ft. (= 0.60 × 0.60 metre); the length of the iron edge is 6.56 ft. (= 2 metres). The depth then of the film of water passing over it when it discharges 6.11 cb. ft. per second, will be 5.10 inches. Fig. 53 gives an elevation of the cast-iron sluice on the

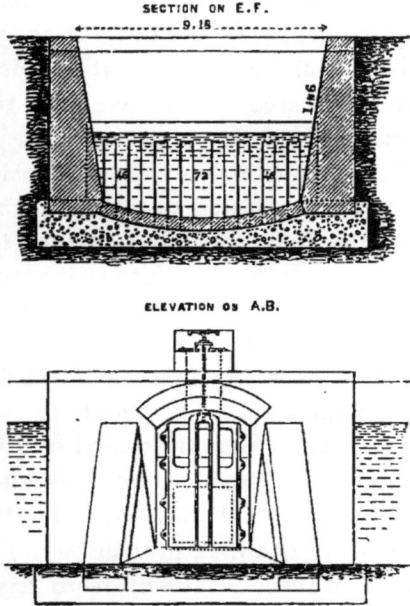

SECTION ON E.F.

ELEVATION ON A.B.

Figs. 52 and 53.

side of the main canal, with the actuating screw, wing-

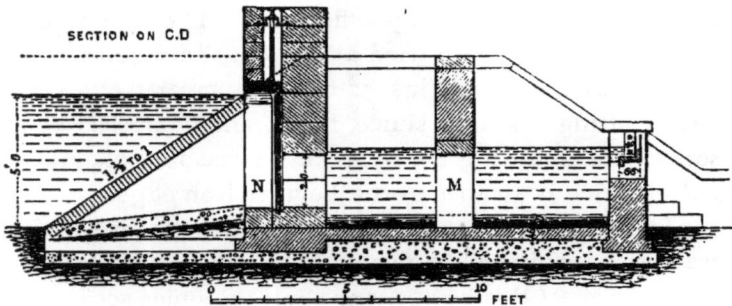

SECTION ON C.D

Fig. 54.

walls, &c., and it is shown in vertical section at N, Fig.

54. The plan, Fig. 55, shows the position of the head of the lifting screw in the coping. The cross wall, with

PLAN.

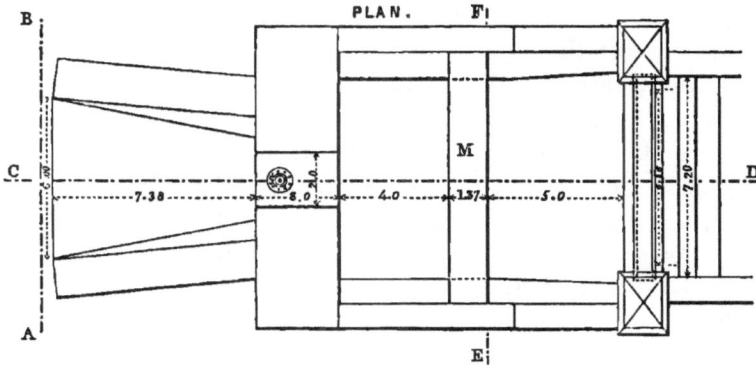

Fig. 55.

its masonry grating, is shown at M in the longitudinal section, Fig. 54, and also in the transverse section through EF, Fig. 52, and again, in the general plan at M.

The cast-iron overfall is shown in section, at an enlarged scale, in Fig. 56, and the upper step on which the water tumbles, all three of which are shown in Figs. 54 and 55. So long as perfect reliance can be placed on the honesty of the guards, the distribution will be effected with great regularity.

The action of the instrument

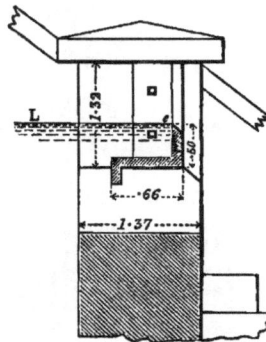

Fig. 56.

for measuring the velocity of rivers, called Pitot's tube, helps to explain some of the subsequent Practical Examples. It is shown in Fig. 57, and consists of a glass tube bent at a right angle, and having at one end a bell-mouth which is immersed in the current horizontally, and turned so as to face up stream, the other end being above the surface and vertical. It is found that the water immediately rises in the vertical part, and it must continue to rise until the column produces an outward pressure at the bell mouth equal and opposite to that caused by the motion of the stream. Oscillations are checked by the bulb on the vertical stem and by a diaphram with a small orifice placed across the mouth. If, therefore, h represent, in feet, the height of the vertical column above the surface, we have the velocity of the stream in ft. per sec. expressed by $v = 8.024 \sqrt{h}$; suppose then it were desired to graduate the tube so that the several num-

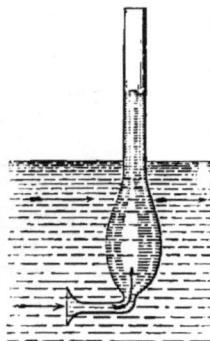

Fig. 57.

bers on its scale should represent velocity of the stream in miles per hour, we have for one mile per hour (since the multiplier $\frac{22}{15}$, or 1.466, alters miles per hour into feet per second)—

$$\left(\frac{22}{15}\right)^2 \times \frac{1}{64.4} \times 12 = 0.4 \text{ inches};$$

for one mile per hour, and for other rates, as follows :—

Miles per Hour,	1	2	3	4	5	6	7	8	9
Inches from the sur-face at which the num-bers 1, 2, 3 are to be placed,	0.4	1.6	3.6	6.4	10	14.5	19.7	25.8	32.7

It is only necessary that the upper part of the vertical tube be of glass, the lower part may be thin copper-plate or other suitable material.

It was by the use of this hydrometer that Pitot overthrew the theory of the old Italian hydraulicians—that the velocity of the several fluid threads in a river increased as the square root of the depth from the surface, and proved, on the contrary, that the velocity diminished from the surface to the bed, as will be mentioned further on.

In Figs. 58, 59 are shown Mr. Ramsbottom's excellent apparatus for filling the tenders of locomotive engines with water while running. It consists of an open trough of water, fixed longitudinally between the rails at about the rail level ; and a dip-pipe or scoop attached to the bottom of the tender, with its lower end curved forwards and dipping into the water of the trough, so as to scoop up the water and deliver it into the tender tank whilst running along. A part longitudinal section of

Fig. 58.

the tender and trough, and part elevation on the right hand, are given in Fig. 58, and a transverse section in Fig. 59.

The water trough A, A, of cast-iron, 18 inches wide at the top, and 6 inches deep, is laid upon the sleepers between the rails, at such a level that, when full of water, the surface is two inches above the level of the rails, its depth being 5 inches. The scoop B (the same letters have the same reference in each Figure), for raising the water from the trough, is of brass, with an orifice 10 inches wide by 2 inches high ; when lowered for dipping into the trough, it has its bottom edge just level with

M 2

the rails and immersed two inches in the water. The
water entering the scoop B is forced up the delivery-pipe
C, which discharges it into the tender tank, being turned
over at the top so as to prevent the water from splashing
over. The scoop is carried on a transverse centre bearing D, and when not in use
is tilted up by the balance-
weight E, Fig. 59, clear of
the ground, as shown by
dotted lines, Fig. 58 ; for dipping into the water trough
it is depressed by means of
the handle and rod, F, from
the foot-plate, which requires
to be held by the engine man
as long as the scoop has to

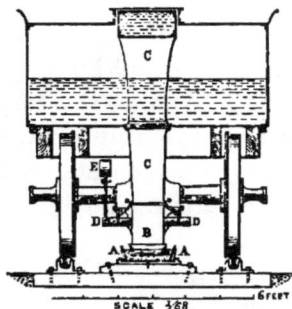

Fig. 59.

be kept down. At N is a fixed strong rod supporting
the transverse bearing D, D.

The upper end of the scoop B is shaped to the form
of a circular arc, as is also the bottom of the fixed delivery-pipe C, so that the scoop forms a continuous prolongation to the pipe when in the position for raising
water. The limit to which the scoop is depressed by the
handle F is adjusted accurately by set screws, which act
as a stop, and prevent the bottom edge of the scoop being
depressed below the fixed working level. The orifice of
the scoop is formed with its edges bevilled off sharp, to
diminish the splashing, and the top edge is carried for-
ward 2 or 3 inches and turned up with the same object.

The principle of action of this apparatus consists in
taking advantage of the height to which water rises in a
tube, when a given velocity is imparted to it on entering
the bottom of the tube—the converse operation being
carried out in this case, the water being stationary, and
the tube moving through it at the given velocity.

The theoretical height, without allowing for friction, &c., is that from which a heavy body has to fall in order to acquire the same velocity as that with which the water enters the tube. Hence, since a velocity of 32.2 feet per second is acquired by falling freely through 16.1 feet vertical, a velocity of 32.25 feet per second, or 22 miles per hour, would raise the water 16.24 feet: and other velocities being proportional to the square root of the height, a velocity of 30 miles per hour would raise the water 30 feet very nearly (a convenient number for reference), and 15 miles per hour would raise the water 7½ feet; half the velocity giving one quarter of the height.

The following Table gives, in the first column, the number of miles per hour at which the train may be advancing; in the second, the equivalent number of feet per second, and the third, the height in feet through which a body must fall, from a state of rest, to acquire that velocity by the action of gravity. The second column is obtained from the first by multiplying the miles per hour by the number 1.466. The third column is the number in the second divided by 8 and the quotient squared:—

Miles per Hour.	Equivalent Feet per Second.	Height fallen vertically to acquire this Speed.
7½ miles.	11.99 ft.	1.888 ft.
15	21.99	7.56
20	29.32	13.43
22	32.25	16.24
25	36.65	20.98
28	41.05	26.32
30	43.98	30.25
35	51.31	41.09
40	58.64	53.73
45	65.97	67.99
50	73.30	83.90
60	87.96	120.78

In the present apparatus the height that the water is lifted is $7\frac{1}{2}$ feet from the level in the trough to the top of the delivery pipe in the tender, which requires a velocity of 15 miles per hour; and this is confirmed by the results of experiments with the apparatus : for at a speed of 15 miles per hour the water is picked up from the trough by the scoop and raised to the top of the delivery pipe, and is maintained at that height whilst running through the trough, without being discharged into the tender.

The maximum quantity of water that the apparatus is capable of lifting is the cubical content of the channel scooped out of the water by the mouth of the scoop in passing through the entire length of the trough : this measures 10 inches wide by 2 inches deep below the surface of the water in the trough, and 441 yards in length, amounting to $\left(\dfrac{20}{144} \times 441 \times 3\right) \times \dfrac{100}{16} = 1148$ gallons, or 5 tons of water. The maximum result in raising water with the apparatus is found to be at a speed of about 35 miles per hour, when the quantity raised amounts to as much as the above theoretical total : so that in order to allow for the percentage of loss that must unavoidably take place, it is requisite to measure the effective area of the scoop at nearly the outside of the metal, which is $\frac{1}{4}$ inch thick and feather-edged outwards, making the orifice slightly bell-mouthed and measuring at the outside $10\frac{1}{2}$ inches by $2\frac{1}{4}$ inches; this gives 1356 gallons for the extreme theoretical quantity.

The result of a series of experiments at different speeds is that at

15 miles per hour, the total delivery is = 0 gals.

22	,,	,,	,,	= 1060	,,
33	,,	,,	,,	= 1080	,,
41	,,	,,	,,	= 1150	,,
50	,,	,,	,,	= 1070	,,

Hence it appears that the variation in the quantity of water delivered is very slight at any speed above 22 miles per hour, at which nearly the full delivery is obtained; the greater velocity with which the water enters at the higher speeds being counterbalanced by the reduction in the total time of action whilst the scoop is traversing the fixed length of the trough.

Mr. Ramsbottom was led to the invention of this apparatus on the occasion of having to provide for the accelerated working of the Irish mail, which has now to be run through from Chester to Holyhead, a distance of $84\frac{3}{4}$ miles, without stopping, in 2 hours and 5 minutes. This necessitated either an increase in the size of the tender tanks beyond the largest size previously used, containing 2000 gallons; or else required the alternative of taking water half-way, at Conway, either by stopping the train for the purpose, or by picking up the water whilst running. A supply of 2400 gallons is found requisite for this journey in rough weather; and, although 1800 to 1900 gallons only are consumed in fair weather, it is necessary to be always provided for the larger supply, on account of the very exposed position of the greater portion of the line, which causes the train to be liable to great increase of resistance from the high winds frequently encountered. An increase of the tender tanks beyond the present size of 2000 gallons would have involved an objectionable increase of weight in construction, and alteration in the standard sizes of wheels and axles, &c., for tenders; and would have also caused a waste of locomotive power in dragging the extra load along the line. By this plan of picking up 1000 gallons at the half-way point where the water trough is fixed, the necessity for a tender larger than the previous size of 1500 gallons is avoided, effecting a reduction in load carried equivalent to another carriage of the train.

For further details reference is made to the description by the inventor in the "Proceedings of the Institution of Mechanical Engineers" for 1861, from which the above has been selected.

The tubes of the Britannia Bridge were constructed on the edge of the shore of the straits near the site of the bridge, and from thence floated to their destination on eight pontoons or barges specially constructed, and arranged in two groups of four each near the ends of the tube which they sustained, to the foot of the abutments, whence they were afterwards raised vertically to their present final position. It was determined to select a tide which would give not more than a maximum relative speed of the current of 9 feet per second; this was too high a rate at which to permit such a mass— more than a thousand tons—to move, and ultimately be arrested at a given position, the limit intended being one foot per second. It became, therefore, a point of importance to ascertain the resistance which the guide cables, passing through stoppers on the pontoons, would have to sustain, and so provide them of sufficient strength, and place ample power at the capstans to check the force of the tide, which was the only motive power employed. The vertical area of the eight pontoons was 400 feet, and upon the principle that the pressure of a current of water per square foot is equal to the weight of a column of water one square foot in base, and of a height equal to that in which a body, falling freely by gravity, would acquire the velocity of the current. Now from $v^2 = 2gh$ we have $h = \dfrac{v^2}{2g}$, and the pressure, P, per square foot, will therefore be, $P = \dfrac{v^2}{2g} \times w$. In which w, the weight of a cubic foot of water, is either $62\frac{1}{2}$ lbs. or 64 lbs., according as it is fresh or salt water; in this example we must take the latter figure. Hence $P = \dfrac{9^2}{64.4} \times 64$. If we take the

divisor and multiplier as equal we have this approximate rule. The pressure of a current of sea water against a vertical surface of one square foot is equal in lbs. to the square of the velocity in feet per second.

For fresh water we have $\dfrac{62.5}{64}$ or $\dfrac{1000}{1024}$ instead of unity. Hence for 1 ft. per second we have a pressure of 1 lb. per square foot, and in the case of 9 feet per second, 81 lbs. on the same area, a result about $1\frac{1}{4}$ per cent. too small. More accurately we have $\dfrac{v^2}{2g} = \dfrac{81}{64.4} = 1.258$ ft., being the altitude from which a body would have to fall freely by gravity to acquire the velocity of 9 ft. per second. And $1.258 \times 64.04 = 82.29$ lbs., which, multiplied by the area and reduced to tons, gives $\dfrac{82.29 \times 400}{2240} = 14.7$ tons. The strength of cables in tons is generally estimated by the square of the circumference in inches divided by 10. The result being about half the breaking weight. Two 12 inch cables were employed for the guide lines.

The single acting Cornish pumping engine, not having any crank axle or other revolving parts on which to key eccentrics for working the valves, is actuated by a contrivance called the "Cataract;" the name, however well applied to the original form, has no connexion with that now used. It is shown in a vertical central section in Fig. 60, in which G is a circular cast-iron tank bolted down on a floor placed below the foot-plate for the engineman; in this is fitted centrally a cylinder *a* in which works a circular plunger, *b*, shown in section in dark lines, guided in the vertical by double glands; on the right-hand side the cylinder opens into a small box cast in one piece with it, having fitted,

in the bottom, a valve, *c*, opening inwards—of the same
construction as that in the lower part of Fig. 43—in the
top it has a cover bolted on with a conical hole in which
fits a conical plug, *d*, at the end of a rod, jointed at *l* to
another vertical rod terminating above in a screwed end
with an adjusting
wheel-handle so that
the plug may be
opened or closed by
the engineman to any
extent required, and
kept fixed in that
position. The plung-
er is worked by the
rod *g*, forked at each
end and connected to
to it below by an eye
bolt which passes wa-
ter tight through the
bottom and is held by
a nut and screw on
the outside: at the

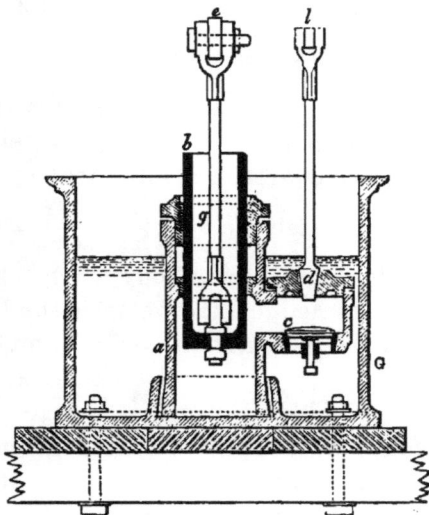

Fig. 60.

upper end it is united by a pin to a lever shown in end
view at *e*. The conical plug *d* and valve *c* are shown
closed, the plunger at half stroke. The action in regulating the number of strokes per minute is as follows :—
The piston descends under steam pressure, let in above
it only, as it is single acting, and a rod from the main
beam descends with it, having a tappet or projection
which strikes the opposite end of the lever *e* and thus raises
the plunger ; the water in the cistern G follows in to supply the partial vacuum thus formed, raising the valve *c*,
which falls when the plunger *b* has ceased to rise.
Now if the plug *d* were quite closed *b* could not descend
at all, and the rate at which it can do so is regulated by

raising *d* more or less, and so the water which has entered through *c* is discharged through the annular opening around *d*. When the plunger has reached the lowest point, the lever strikes a detent, which frees a weight by which, and not by an eccentric, the steam valve is opened; thus the number of strokes per minute depends upon the length of time in which the plunger *b* is descending: from 2 to 12 strokes per minute are about the limits.

A branch of the Midland Railway (Ireland) crosses the Royal Canal at 2 feet above the water surface, and to provide for the passage of barges, the engineer, Mr. J. Price, has constructed a vertically lifting bridge; this, when raised sufficiently, is, after the traffic has passed under, let down with a regulated descent by four plungers, one over each angle of the abutment; these work in cast-iron cylinders firmly bolted to the masonry, having a small orifice opening into the water of the canal; on the rise of the plungers the water follows them and they cannot descend faster than it can escape. The principle has been applied in numerous other instances.

The native troops in India are accustomed to relieve guard, on the sinking of a perforated metallic cup in a vase of water. As a converse of this, the ancients, instead of a sand-glass, employed a cistern, from which the water trickled through a small hole at the bottom, under the name of a *Clepsydra* or *water-clock*, to measure time. In a cylinder the rate of descent of the surface diminishes, as the level of the water is successively lowered. To obtain an uniform descent of the water, it would be necessary to adopt the figure of a conoid of the parabolic kind, each

Fig. 61.

circular section of which is proportional to the square root of the corresponding altitude, and since (if *d* and *h*

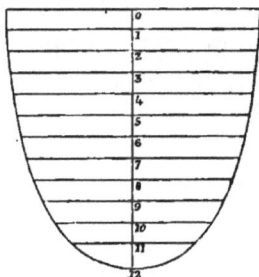

be corresponding diameters and heights) d^2 is proportional to $\sqrt{h}$, we have d^4 proportional to h. Let h at the commencement be 24 ft., and d = 13.28 ft., also let the uniform rate of descent be 1 ft. per hour to find x the diameter at the end of the fourth hour, then

$$24 : 20 :: 13.28^4 : \frac{20}{24} \times 13.28^4 = 25917.5 = x^4,$$

whence x = 12.688 ft. At the end of the twentieth hour the diameter in like manner should be 8.46 ft. To compute the diameter of the orifice we have the velocity of descent at the surface to that at the orifice inversely as their areas, that at the surface being 1 ft. per hour, and at the bottom $8.024 \sqrt{24}$ = 39.2 ft. per second, which we must multiply by 3600 to bring them to the same unit of time, $3600 \times 39.2 : 1 :: 13.28^2 : z^2$, or, taking the square roots $60 \times 6.261 : 1 :: 13.28 : z = 0.424$ inches. A conoid of such dimensions would, as represented in Fig. 61, therefore, answer correctly as a Clepsydra, the equable subsidence of a float marking the series of twenty-four hours in a natural day. This float being fastened to a thread wound about a cylindrical barrel, one foot in circumference, would carry the index of a dial regularly round.

108. *Examples on Weirs.*—Some older writers, as Hutton, &c., give a geometrical construction to represent the discharge through notches and weirs different from that in pp. 49 and 52—namely, if ACDB, Fig. 62, be the transverse section of the sheet of water flowing over, then, with either side, as AC, for axis, and A the vertex, drawing a parabola passing through D ; the volume discharged is equal to ACDSA multiplied into the line representing, at the same scale, the velocity in the lowest line CD.

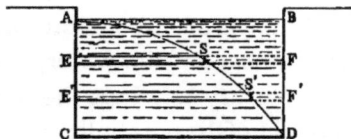

Fig. 62.

Let ACDB represent in elevation a rectangular

notch through which water is flowing from a reservoir
maintained at the constant level AB, then the quantity
discharged will be ⅔rds of the quantity flowing through
an equal orifice placed at the whole depth AC in the
same time. For if we suppose the transverse section to
be divided into an indefinite number of elementary
rectangles, as EF, E'F', CD, by equidistant parallel lines,
then the volumes of water discharged through each of
these equal rectangles will be proportional to the square
roots of the depths, that is, to $\sqrt{AE}$, $\sqrt{AE'}$, $\sqrt{AC}$, and
the sum of all these discharges will be the total dis-
charge through the whole area ACDB. But the discharge through each rectangle, as EF, with the velocity
due to its particular depth, will be equal to that of a rectangle of less width, as, suppose, ES, at the same depth,
provided its velocity is increased inversely as the width
is diminished. Now, if through A as vertex, a parabola
be drawn through D, the vertical AC being the axis, and
if we terminate these lesser rectangles in the curve so
drawn, then the velocity in each will be identical and
equal to that of the lowest lamina, for the volume discharged at CD is to that discharged at any other depth
EF, as $\sqrt{AC}$ is to $\sqrt{AE}$, that is, as the line CD, or its
equal EF, is to the line ES; and the velocity in the rectangle ES is to that in the rectangle EF, when they have
equal discharges, as EF to ES.

The total discharge through the notch, in one second,
ACDB, is therefore equal to the volume of a prism
having the parabolic segment ACD S'SA as base and
the velocity at CD for its height, and as the parabola
is ⅔rds of the rectangle, the discharge will be ⅔rds of
an equal area placed at the depth CD.

Let us, as in p. 119, exhibit the effect of the different values of m on the quantity discharged per second, namely, 0.60, 0.665, and between 0.662 and
0.595.—*Vide* § 77, pp. 93-94.

Thus, suppose an overfall of 1 ft. in width, having a depth of 1 ft. passing over: required the discharge in one second; the formula $\frac{2}{3} mlH \sqrt{2gH}$ then becomes $\frac{2}{3} \times m$ × 8.024, or m × 5.35.

1. $m =$ (§ 64, p. 75) 0.60, value of $Q =$ 3.21 cb. ft.
2. $m =$ (§ 64, p. 75) 0.665 „ = 3.558
3. $m =$ (§ 82, p. 100) 0.66654 „ = 3.566
4. $m =$ (§ 64, p. 77) 0.595 „ = 3.183

In the following questions it is intended to show the effect of the function of the head or charge $H\sqrt{H}$, which occurs in the formula for the discharge over weirs. A certain length is taken, and the discharge with a given head determined, and then this discharge being increased by a given quantity (XXV.), the corresponding increase of H is determined. In the same, the discharge being doubled, it is sought (XXVI.) to determine the relative increase of the value of H.

(XXIV.) Calculate the discharge over a weir 1100 ft. long, the depth from the surface of still water to the crest of the weir being 0.75 ft., using 0.665 for the value of m, (as in second case above), we have 8.024 × $\frac{2}{3}$×0.665=3.558. Hence 3.558 × 1100 × 0.75 × 0.866 = 2541.3 cb. ft. per sec., as $\sqrt{0.75} =$ 0.866. In Beardmore's Table II. we find the discharge for 1 ft. of length of weir with 0.75 ft. head, 138.88 cb. ft., and this multiplied by 1100, gives 152768; but as all his Tables are calculated for the discharge per minute, dividing 152768 by 60, we obtain 2546.13, differing from that calculated above by 4.83; the coefficient used by Beardmore being 0.6665 (§ 82, p. 100), giving 3.566, instead of 3.558 used above.

(XXV.) To what height upon the crest would the water rise if the discharge was increased to 3000 cb. ft. per sec. ? We have from these data,

$$3.558 \times 1100 \sqrt{H^3} = 3000.$$

Hence

$$H = \sqrt[3]{\left(\frac{3000}{3.558 \times 1100}\right)^2} = \sqrt[3]{(0.766)^2} = \sqrt[3]{0.587} = 0.837,$$

being an increase of 0.837 − 0.75 = 0.087 ft., the increase of H being only 11.6 per cent., and that of Q being 17.9 per cent.

The least laborious method of finding cube roots, when no table of logarithms is at hand, is the following:—Assume a number whose cube is nearly equal to the given number; then, as twice this cube, plus the given number, is to twice the given number, plus the assumed cube, so is the assumed root to the true; in this case, for $\sqrt[3]{0.587}$ first assume 0.8, which gives 0.512; secondly, assume 0.84, which gives 0.593. Hence

$$\overline{2 \times 0.593 + 0.587} : \overline{2 \times 0.587 + 0.593} :: 0.84 :: 0.837.$$

And by logarithms we have log. of 0.587 = $\overline{1}$.7686381, which, divided by 3, gives 3) $\overline{3}$.7686381

$\overline{1}$.9228793 answering to 0.83728.

(XXVI.) If the discharge in (XXIV.) had been doubled, calculate the depth of water flowing over the crest. The average discharge in (XXIV.) being doubled, gives (2 × 2544 =) 5088 cb. ft. per second on a length of 1100 ft. Hence

$$H = \sqrt[3]{\left(\frac{5088}{3.558 \times 1100}\right)^2} = \sqrt[3]{1.69}, \text{ and}$$

log. of 1.69 = 0.2278867, which, divided by 3, gives 0.0759622, answering to 1.19131; deducting 0.75, we have 0.44 ft. for the rise, to be added to the first supposed 0.75, in order to obtain a double discharge, so that, instead of

1.50, i. e. twice the original head, we have but 1.19 ft. on the crest of the weir for twice the original discharge; it is, in fact, evident that 0.75 is multiplied by $\sqrt[3]{2^2} = 1.5866$, instead of by 2.

If the length of the weir in (XXIV.) had been reduced one-half, namely, to 550 ft., calculate the head to which the water would rise upon the crest, the discharge being the same, namely, 2544 cb. ft. per sec. We have now

$$Q = 2544 = 3.558 \times 550 \times H^{\frac{3}{2}}.$$

Hence

$$H = \left(\frac{2544}{1956.9} \right)^{\frac{2}{3}} = 1.191 \text{ ft.}$$

(XXVII.) The construction of the weir at Killaloe ("Selection of Specifications") has the peculiarity of not being level on part of the crest. The inclination being 1 in 214, and the rise 1.5 ft., the length with that slope must be $1.5 \times 214 = 321$ ft.; we have therefore as the weir is 1100 ft. long, 779 ft. for the level portion, and 321 ft. at an inclination of 1 in 214. Calculate the total quantity discharged over this weir when the depth of water on the level part is 1.8 ft., so as to have 0.3 ft. on the highest part of the crest at the west abutment. If then we divide this sloping part into eight lengths, of 40 ft. each, and calculate the discharge over each length with a head equal to the arithmetic mean of the head at each extremity of the 40 ft. lengths, the discharges will be sufficiently near the truth. The increase of depth on each 40 ft. is evidently $\frac{40}{214}$ ft., equal to 0.18691 ft., and as the depth over the highest point at the west abutment is, by the terms of the question, 0.3 ft., the mean depth for the first 40 ft. is

$$\frac{0.3 + 0.3 + 0.18691}{2} = 0.393455 \text{ ft.;}$$

to obtain the second, third, &c., we have but to add to this successively 0.18691, and consequently obtain the following numbers:—0.393, 0.580, 0.767, 0.954, 1.141, 1.328, 1.702 1.795; which, being multiplied by their respective square roots, give 0.2468, 0.442, 0.672, 0.932, 1.219, 1.530, 1.864, 2.220.

Hence the eight several discharges through the 40 ft. lengths are found by multiplying the common part of the formula (§ 55) $\frac{2}{3} m, l, H \sqrt{H} . \sqrt{2g}$, that is, 3.558×40 $= 142.3$ into the values of $H \sqrt{H}$ given above, and, adding these, we have the total discharge over the sloping part of this weir 1299 cb. ft. per sec. And for the length of 780 ft. of level crest with 1.8 ft. head, we have 6700 cb. ft. per sec. Hence the total discharge is 7999 cb. ft. per sec. As $8 \times 40 = 320$ ft., and the length of sloping portion is 321 ft., we must add one foot to 779, the length of the level portion.

(XXVIII.) In the weirs on the Shannon constructed by the Commissioners, it was requisite that salmon-gaps should be constructed, so that the fish be able to migrate up stream at the weirs during such periods as might not afford sufficient depth of water if the whole quantity were uniformly distributed over the total length of the weir. These were 10 feet wide, and the crest 1.5 ft. below that of the weir. Calculate the quantity flowing down three of these salmon-gaps, the water on the level part of the crest being 0.6 ft. deep. Here

$$H = 1.5 + 0.6 = 2.1, \text{ and}$$

$$3Q = 3 \times 3.558 \times 10 \times 2.1 \sqrt{2.1} = 324.8 \text{ cb. ft.}$$

(XXIX.) A feeder or water-course along the side of a valley is required to be augmented by the streams and springs above its level. It is required to determine their total volume. For this purpose the several courses are dammed up at convenient and suitable places, and a narrow board provided, in which is cut an opening for the overfall 1 ft. long, and 0.5 feet deep; it being reasonably surmised that this would be sufficient to gauge the largest of the streams; and another piece was prepared that, when attached to the former, would reduce the length to 0.5 ft. for the smaller. Calculate the total quantity delivered by the five following streams and springs:

No. 1, on being dammed up, flowed over the 1 ft. opening 0.37 ft. deep. Hence $Q = 3.558 \times 0.37 \sqrt{0.37} = 0.8$ cb. ft.

No. 2, at 0.5 ft. in length of overfall, rose to 0.41 ft. in depth. Hence $Q = 3.558 \times 0.5 \times 0.41 \sqrt{0.41} = 0.467$ cb. ft. per second.

No. 3, at 1 ft. length, was 0.29 ft. high on the overfall, and

$$Q = 3.558 \times 0.29 \sqrt{0.29} = 0.555 \text{ cb. ft. per sec.}$$

No. 4, at 0.5 in length, rose 0.19 ft. Hence we have

$$Q = 0.5 \times 3.21 \times 0.19 \sqrt{0.19} = 0.133 \text{ cb. ft. per sec.}$$

No. 5, being a small spring, was not measured by the overfall; but being banked up, a pipe, 0.0416 ft. in diameter, was let through the dam, and when the surface had become stationary, and consequently the discharge through the pipe equal to the supply from the spring, it was gauged into a vessel marked for 1 and 2, &c., imperial gallons; the time required to reach the former was 32 seconds. Hence the spring gave 0.005 cb. ft. per sec., as 6.25 gallons make one cubic foot.

The total quantity, therefore, received by the aqueduct from the lateral springs and streams above its level amounted to 1.96 cb. ft. per second.

(xxx.) On the Manchester water-works weirs are constructed across some of the lateral mountain streams which supply the reservoirs, so that the higher velocity which the water has when flowing over at the greater depths may separate the turbid water, unfit for the town supply, from the clear. In heavy or sudden rains these streams bring down very rapidly water discoloured by peat and earth, and unfit for domestic use ; but in fine weather the quantity is much reduced, and the water clear and suitable for the mains of the town. The wood engraving represents a transverse section of the water-course which is carried through the masonry of the weir, conveying clear water from other streams, across the valley in which the weir is placed, and so serving as an aqueduct ; at the top this is open, and when the water flows over at a small depth, that is, when it is clear, it falls into the channel, Fig. 63, and is con-

Fig. 63.

veyed by it eventually into the main which supplies the town; but if it rise and discharge a greater body of water, the increased velocity projects it beyond the edge of the opening, Fig. 64, and it thus passes over the longitudinal opening, and flows down to the compensation reser-

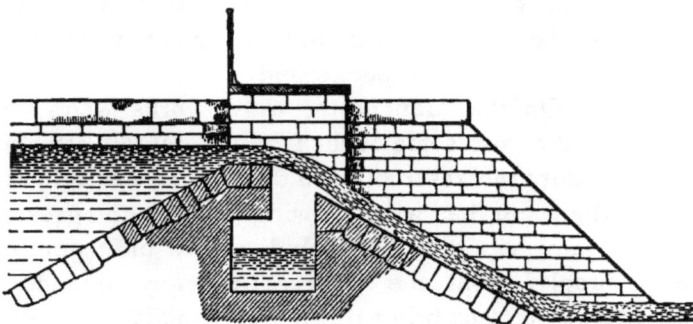

Fig. 64.

voir for the supply of water to the mills situated on the river.

By referring to § 8, we find the means of calculating the curve of any issuing jet of water. But in this case we have a different velocity, and therefore a different parabola for every lamina into which we may suppose the water divided. Fig. 65 represents the different paths taken by each, that for the mean velocity at $\frac{4}{9}$ths of the depth being drawn in a full line ; hence those above will

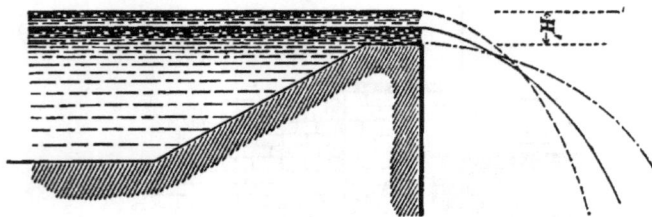

Fig. 65.

tend to depress the curve, and those below, on the contrary, to carry it more up towards the horizontal line ; we may therefore suppose the whole sheet of water to be carried out in a curve at top and bottom parallel to that of the mean velocity, Fig. 66. If therefore we put H_1 for the depth of the water flowing over the weir, the mean velocity being $\frac{2}{3}$rds of that at the bottom, we have

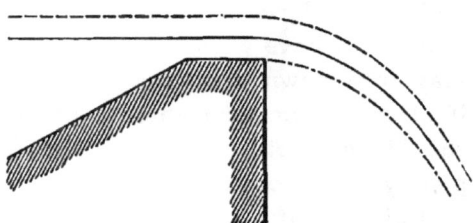

Fig. 66.

$$v = \frac{2}{3} \times 8.024 \times \sqrt{H_1}$$

for this mean velocity, and the curve taken by the lowest lamina is that due to a head $\frac{4}{9} H_1$, for if in the expression (§ 48, p. 52)

$$z' = \frac{4}{9}\left(\frac{H \sqrt{H} - h \sqrt{h}}{H - h}\right)^2,$$

we put $h = 0$, the resulting value of z' is $\frac{4}{9} H$. Now in Fig. 67 let $x = 1$ ft., and $y = 0.83$ ft.; hence, from § 8, p. 12,

$$\frac{4}{9} H_1 = \frac{y^2}{4x} = \frac{0.83^2}{4} = 0.1722 \; \therefore \; H_1 = 9 \times 0.1722 \div 4 = 0.3874 \text{ ft.}$$

So then, when the water flowing over has a depth at or greater than 0.3874 ft., it is carried completely over the longitudinal opening. We must, then, gauge the stream in wet seasons, and so proportion x to y that the volume of water, from the head necessary to discharge it, have velocity sufficient to pass over the opening mn; at lesser depths it strikes against

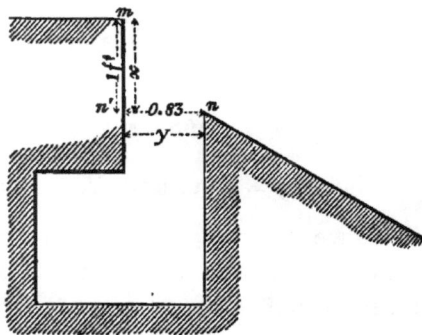

Fig. 67.

the point, and in part enters the clear water-channel, and in part flows over the weir; for this reason it is necessary to have a cover of timber, that the attendant may turn down upon the opening during such period if, at the commencement or end of a flood, the water should be turbid at such a depth as would not completely pass over the opening *m n*.

Calculate at what depth the water all flows in. If we suppose in Fig. 68 that $nr = H_1$, which we may do, though

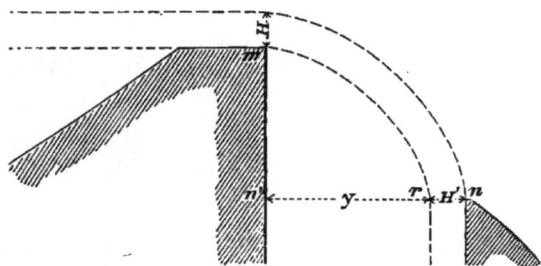

Fig. 68.

it be not normal to the axis of the sheet of water, then $y + H_1 = 0.83$ and $H_1 = 0.83 - y$, also $\frac{4}{9} H_1 = \frac{y^2}{4}$ in this substitute for H_1 its value above, we have

$$\frac{4}{9}(0.83 - y) = \frac{y^2}{4} \text{ and } \frac{4}{9} \times 0.83 = \frac{y^2}{4} + \frac{4}{9} y ;$$

$$\text{or } \sqrt{2.266} = y + \frac{8}{9}, \text{ or } 1.5 - 0.888 = y ;$$

hence $y = 0.612$ and $H_1 = 0.833 - 0.612 = 0.221$ ft.

Thus we see that, when the opening is constructed so that $x = 1$ ft., and $n' n = 0.83$ ft., a depth of water on the crest of 0.3874 ft., or more, carries all the water over the opening, and a depth of 0.221 ft., or less, admits all.

If, then, we observe in ordinary seasons a stream discharging 26.6 cb. feet per sec., the water then being clear,

and the most convenient length of the crest of the overfall being 60 ft.,—we may, having selected some convenient depth as $n_1 m$, so adjust the opening mn that the whole
of the clear water may fall into it. As a first step, we must
calculate H_1, for the length 60 feet, and a discharge of 26.6
cb. ft., if the coefficient m be taken equal to 0.6665 (§82); we
have, therefore, $H_1 = \sqrt[3]{\left(\dfrac{26.6}{3.556 \times 60}\right)^2} = 0.25$ (§71), and the
vertical depth of n or of n_1 (Fig. 68) below the crest at m
being given, we may calculate, first, the value of y, that
is $n_1 n$, so that the curve of the *upper* surface of the sheet
of water flowing over the crest may fall within the point
n, and the whole stream be carried down the clear water
aqueduct. Let mn_1 be taken equal to 1 ft., then from

$$y^2 = \frac{4v^2 x}{2g} \text{ (§ 8)}, \text{ we have } y = \frac{2v\sqrt{x}}{\sqrt{2g}}, \text{ and}$$

substituting for v its value in this particular case where
$H_1 = 0.25$, we have $v = \dfrac{2}{3}\sqrt{2gH_1} = \dfrac{2}{3} \times 8.024\sqrt{0.25}$ (§ 48),
and also for x its value, which is $mn_1 + H_1 = 1.25$ ft.
Hence

$$y = \frac{2 \times \dfrac{2}{3} \times 8.024\sqrt{0.3125}}{8.024} = \frac{4}{3} \times 0.559 = 0.745 \text{ ft.}$$

And secondly, we may calculate at what amount of discharge and head H_1 the curve of the *lower* parabola of
the sheet of water will pass completely over the opening
mn, and so the stream, now turbid, be all carried over
the clear water aqueduct into the settling and compensation reservoirs. Call the sought depth D, and as x is

now 1 ft., we have $y = 0.745 = \dfrac{2 \times \dfrac{2}{3} \times 8.024\sqrt{D}}{8.024} = \dfrac{4}{3}\sqrt{D}$ and

$D = \left(\frac{3}{4}\,0.745\right)^2 = 0.31248$, the discharge being about 37 cb. ft. per sec.

(XXXI.) To determine generally the relation between the length and depth of the water on a weir having the same discharge, put

$$\frac{2}{3} \times m \times l_1 \times h_1 \ \sqrt{h_1} = \frac{2}{3}\, m \times l_2 \times h_2 \ \sqrt{h_2} \,;$$

hence,

$$h_1^{\frac{3}{2}} : h_2^{\frac{3}{2}} : l_2 :: l_1,$$

and

$$h_1 : h_2 :: l_2^{\frac{2}{3}} : l_1^{\frac{2}{3}},$$

$$\therefore h_1 = h_2\left(\frac{l_2}{l_1}\right)^{\frac{2}{3}},$$

$$\log h_1 = \log h_2 + \frac{2}{3} (\log l_2 - l_1).$$

Calculate the height to which the water upon a weir 545 ft. long will rise when it is flowing down from another weir higher up upon the same river, whose length is 750 ft., and on which it rises 0.68 ft., it being supposed that no additional supply has been received in the intervening part of the course.

Here $\log l_2 = 2.8750613$
          $\log l_1 = \underline{2.7363965}$
              $0.1386648 \times \dfrac{2}{3} = 0.0924432$

and $\log h_2 = \overline{1}.8325089$
              $\underline{0.0924432}$

              $\overline{1}.9249521$ and $h_1 = 0.84$ ft.

By the above method, namely, by discharging the same quantity of water over weirs of different length and measuring the depths, may be determined experimentally the value of the index of H to which the discharge is proportional, on the supposition that m is constant, and that the discharge is directly as the length, for then

$$Q = \tfrac{2}{3} \cdot m \cdot l_1 \cdot h_1{}^a = \tfrac{2}{3} \cdot m \cdot l_2 \cdot h_2{}^a,$$

and therefore,

$$\left(\frac{h_1}{h_2}\right)^a = \frac{l_2}{l_1}, \text{ or}$$

$$a \,(\log h_1 - \log. h_2) = \log l_2 - \log l_1,$$

and

$$a = \frac{\log l_2 - \log l_1}{\log h_1 - \log h_2}.$$

The following Table is arranged from Series VII., Table X. of J. B. Francis, "Lowell Experiments," already mentioned, § 61, p. 68 :—

TABLE *showing the Results of Experiments to determine the Index of H.*

	Total length of Weir in feet.	Or, very nearly,	Depth on crest of Weir in feet.	· Average Index.
1	16.980	17.0	0.51837	
2	13.978	14.0	0.59514	
3	10.489	10.5	0.72733	1.478
4	8.489	8.5	0.83614	
5	6.987	7.0	0.95882	
6	5.487	5.5	1.13087	

(XXXII.) In the construction of reservoirs it is necessary to have a weir whose crest is on the level of the intended top-water line, with reference to which line the height of the embankment and of the puddle-wall must also be designed. Its length must be such that the water of a maximum rain-fall shall not rise on it above a certain height. We may take the greatest available rain-fall at 2 inches in 24 hours; this depth must be multiplied into the area of district which drains into the reservoir. We thus have, first, the total volume of water; and secondly, supposing the rain to have fallen at a uniform rate during the 24 hours, or at least to have been delivered by the water-courses into the reservoir at a uniform rate, we thence obtain the quantity per minute or per second which this weir must discharge. We then assign a certain depth upon the crest, to which the water must be limited, and consequently, from the depth H and discharge Q we obtain L.

Thus, suppose the area of the rain-basin or district draining into the reservoir were 6536 acres, and the maximum depth of rain in 24 hours to be 2 inches, we reduce both to the same unit of feet. The acre contains 10 square chains of 66 feet each, $66^2 \times 10 = 43560$ sq. ft., and 2 inches = 0.1666' ft.: hence, $43560 \times 6536 \times 0.1666'$ = 47,451,360 cb. ft. in 24 hours, which, reducing to seconds, we have $24 \times 60 \times 60 = 86400$; and dividing $47,451,360 \div 86400 = 549.2$ cb. ft. per second, entering the reservoir, the length of weir to discharge this with a rise on the crest of 1.5 ft. is found

$$\frac{549.2}{\frac{2}{3} \times 0.66 \times 1.5 \sqrt{1.5} \times 8.024} = \frac{549.2}{6.48} = 84.75 \text{ ft.}$$

As, however, the valves for discharging the storage would be opened, the rise upon the crest could be readily kept down to one foot.

CHAP. II.—FLOW OF WATER UNDER A VARIABLE HEAD.

109. In § 88 we have the formula $T = \dfrac{2A\sqrt{H}}{mS\sqrt{2g}}$.

(XXXIII.) This has been used to determine the value of m. A tube 1 inch in diameter is filled 9 inches in depth with mercury; at the bottom is an orifice $\frac{1}{20}$ inch in diameter; the observed time of its total discharge was 140 seconds. Solving for m, we have $m = \dfrac{2A\sqrt{H}}{T.S.\sqrt{2g}}$. Changing the measures from inches into feet, we have—

$2A = 0.083^2 \times 0.7854 \times 2 = 0.0109$ sq. ft. and $\sqrt{0.75} = 0.866$ ft.

$$S = \frac{1}{400} \times 0.00545 = 0.0000136 \text{ sq. ft.}$$

And $m = \dfrac{0.0109 \times 0.866}{140 \times 0.0000136 \times 8.024} = \dfrac{0.0094394}{0.0152777} = 0.62.$

From the vortex motion of the fluid at small depths, no formulæ which give the time for *complete* exhaustion are quite exact. Mercury is, probably, less affected by this motion than water, with which a funnel-shaped vortex is formed over the orifice; this drawing in the air renders the discharge irregular, and reduces the orifice, so that the formula for *partial* exhaustion—

$$t = \frac{2A\left(\sqrt{H}-\sqrt{h}\right)}{mS\sqrt{2g}},$$

gives more exact results, as in the following experiment.

(XXXIV.) A prismatic vessel, having a diameter of 5.747 inches, has an orifice of 0.2 inch at the bottom, and its surface is observed to sink from 16 inches to 1 foot of depth in 53 seconds. Transposing as before, we have—

$$m = \frac{2A\left(\sqrt{H} - \sqrt{h}\right)}{t \times S \sqrt{2g}},$$

H being 1.33′ ft., and $h = 1$ ft., the value of $\left(\sqrt{1.33} - \sqrt{1}\right)$ is $1.153 - 1 = 0.153$ ft. The diameter of the vessel being 5.747 inches, or 0.4783 ft., the value of A will be $0.4783^2 \times 0.7854 = 0.1797$ sq. ft.: also $S = 0.0166^2 \times 0.7854 = 0.000218$ sq. ft. Hence—

$$m = \frac{2 \times 0.1797 \times (1.153 - 1 =) 0.153}{53 \times 0.000218 \times 8.024} = \frac{0.055}{0.0927} = 0.60, \ q.\ p.$$

(XXXV.) A prismatic basin, whose horizontal section is a square of 3 ft. in the side, has at the bottom an orifice 0.09 ft. in diameter; it is filled up to a depth of 6 ft. above the centre of the orifice. Calculate the *time* required for the surface to descend 3.5 ft., counting from the moment of opening the orifice. Here $A = 3 \times 3 = 9$ sq. ft., $S = 0.09^2 \times 0.7854 = 0.00636$ sq. ft.; $H = 6$, and $h = 6 - 3.5 = 2.5$, m being 0.61; therefore from the formula—

$$t = \frac{2 \times 9\ (2.449 - 1.581)}{0.61 \times 0.00636 \times 8.024} = \frac{15.624}{0.03113} = 502'' = 8'\ 22''.$$

(XXXVI.) With the same dimensions calculate the time required for the surface to descend 2 ft. Here $h = 6 - 2 = 4$, and $\sqrt{H} - \sqrt{h} = 0.449$ ft.; therefore—

$$t = \frac{18 \times 0.449}{0.03113} = 259.6 = 4'\ 20''.$$

(XXXVII.) Again, suppose the descent of the surface to be 5 ft., calculate the time, $h = 6 - 5 = 1$, and $\sqrt{H} - \sqrt{h} = 1.449$, so that—

$$t = \frac{18 \times 1.449}{0.03113} = \frac{26.082}{0.03113} = 837''.84 = 13'\ 58''.$$

(XXXVIII.) § 91. *Mean Hydraulic Charge.*—Let us suppose in any prismatic vessel receiving no supply, that the head, at the instant of opening the orifice of discharge, was 6 ft. = *H*, and at closing it had decreased to 5 ft. = *h*, calculate the mean constant charge at which, in the same time, the orifice would discharge the same volume of water; the vessel being now, necessarily, supposed to receive that same constant quantity which it discharges with a uniform velocity.

The formula is—

$$H' = \left(\frac{H - h}{2\left(\sqrt{H} - \sqrt{h}\right)} \right)^2 = \left(\frac{6 - 5}{2(2.449 - 2.236)} \right)^2 = \left(\frac{1}{0.426} \right)^2 = 5.508 \text{ ft.}$$

If *h* be taken equal to 4, then $H' = 4.96$; if equal to 3, $H' = 4.376$; if $h = 2$, then $H' = 3.732$; and when $h = 0$, we have $H' = 1.5$.

If in 10″ we observe the surface to fall 2 ft., determine the coefficient of discharge.

If $A = 6$ ft., $S = 0.01$, and $T = 10″$, then H being = 6, and $h = 4$, we have $Q' = 12$ cb. ft., and $Q = 1.2$, per sec., $H' = 4.96$, and $\sqrt{H'} = 2.227$ ft.

Hence—

$$m = \frac{1.2}{0.1 \times 8.024 \times 2.227} = \frac{1.2}{1.787} = 0.67.$$

(XXXIX.) § 92, p. 109. A reservoir, half an acre in area, with sides nearly vertical, so that it may be considered prismatic, receiving a stream which yields 9 cb. ft. per second, discharges through a sluice 4 ft. wide, which is raised 2 ft.; calculate the time required to lower the surface 5 ft., the charge upon the centre of the sluice, when opened, being 10 ft. From the formula given at the end of § 92, we have, substituting the numerical values, $A = 21780$ sq. ft. the acre, being 43560 sq. ft.; $S = 8$ sq. ft., *m* being found 0.70, and $h = 10 - 5 = 5$, also

$q = 9$ sq. ft. per second—

$$t = \frac{2 \times 21780}{(0.7 \times 8 \times 8.024)^2} \left\{ 0.7 \times 8 \times 8.024 \left(\sqrt{10} - \sqrt{5} \right) \right.$$

$$\left. + 2.303 \times 9 \times \log \frac{0.7 \times 8 \times 8.24 \ \sqrt{10} - 9}{0.7 \times 8 \times 8.024 \ \sqrt{5} - 9} \right\}.$$

In this we have $0.7 \times 8 \times 8.024 = 44.9$, and $\sqrt{10} - \sqrt{5}$
$= 3.162 - 2.236 = 0.926$.

Hence—

$$t = \frac{43560}{2016} \left\{ 44.9 \times 0.926 + 20.7 \log 1.455 \right\}$$

$$= 21.607 \left\{ 41.6 + 3.37 \right\} = 972'' = 16', 12''.$$

If q, the constant supply received by the reservoir, had been 20 cb. ft, per second, then—

$$\frac{(44.9 \times 3.162) - 20}{(44.9 \times 2.236) - 20} = \frac{121.97}{80.40} = 1.517,$$

the log. of which is 0.1809856 (in the former case subtracting 9 we had $\frac{132.97}{90.4} = 1.455$, the log. being 0.1628630), and the value of t is now $21.607 \left\{ 41.6 + 2.303 \times 20 \times 0.181 \right\}$ $= 1079'' = 17' \ 59''$ to lower the surface 5 ft.

(XL.) Referring to the latter part of § 92, in order to determine the depth which the surface would descend in a given interval of time, the formula must be arranged so as to separate the factors of $\sqrt{H}$ from $\sqrt{h}$, then transposing, so as to make the left-hand side = 0, we have—

$$t - \frac{2A}{(mS\sqrt{2g})^2} \left\{ mS\sqrt{2g}\sqrt{H} + 2.303 \times q \times \log \left(mS\sqrt{2g}\sqrt{H} - q \right) \right\}$$

$$+ \frac{2A}{(mS\sqrt{2g})^2} \left\{ mS\sqrt{2g}\sqrt{h} + 2.303 \times q \times \log . \left(mS\sqrt{2g}\sqrt{h} - q \right) \right\} = 0.$$

Let us suppose all the letters to have their former values, t being taken at 20 minutes, calculate the value of h—

$$(t=)\ 1200'' - \frac{43560}{2016}\ [44.9 \times 3.162 + 20.73 \times \log 133\} =$$

$$1200 - 4020 = -2820,$$

and thus we have—

$$21.61 \times [44.9\ \sqrt{h} + 20.73 \times \log (44.9\ \sqrt{h} - 9)] - 2820 = 0,$$

when the true value of h is substituted. To further prepare this last expression for the tentative determination of h, we multiply out by 21.61, hence—

$$970.3\ \sqrt{h} + 448 \log (44.9\ \sqrt{h} - 9) - 2820 = 0.$$

If we take at at first—

$\sqrt{h} = 2$, the equation becomes $-\quad 25 = 0$,

$\sqrt{h} = 2.4$ „ „ $+\quad 422 = 0$,

$\sqrt{h} = 2.01$ „ „ $-\quad 14 = 0$,

$\sqrt{h} = 2.1$ „ „ $+\quad 82.7 = 0$,

$\sqrt{h} = 2.03$ „ „ $+\quad 7.44 = 0$,

$\sqrt{h} = 2.023$ „ „ $-\quad 0.1 = 0$, and $h = 4.09$.

The surface, therefore, descends 5.9 feet in 20′.

(XLI.) A pond, whose area is 12000 square feet, has an overfall outlet 3 feet wide, which at the commencement of the discharge has a head of 2.8 feet; calculate the length of time required for the surface to descend 1 foot, it being supposed that no supply is received.

We have then $H = 2.8$, and $h = 2.8 - 1 = 1.8$, the value of m being taken at 0.61.

The formula, § 93—

$$t = \frac{3A}{ml\sqrt{2g}}\left(\frac{1}{\sqrt{h}} - \frac{1}{\sqrt{H}}\right)$$

being put into numbers for this question, we have—

$$t = \frac{3 \times 12000}{0.61 \times 3 \times 8.024}\left(\frac{1}{\sqrt{1.8}} - \frac{1}{\sqrt{2.8}}\right) = 2452\left(\frac{1}{1.34} - \frac{1}{1.673}\right)$$

$$= \frac{2452}{1.34} - \frac{2452}{1.673} = 1830 - 1466 = 364'' = 6'\ 4''.$$

Calculate the time in which the surface descends 0.5 feet. In this case $h = 2.8 - 0.5 = 2.3$, and $\dfrac{1}{\sqrt{2.3}} = \dfrac{1}{1.516}$. Hence—

$$\frac{2452}{1.516} - \frac{2452}{1.673} = 1617 - 1466 = 151'' = 2'\ 31''.$$

Again, if we suppose the depth descended to be 1.5, and all the other quantities remain the same, we shall thus have—

$h = 2.8 - 1.5 = 1.3$, and $\dfrac{1}{\sqrt{1.3}} = \dfrac{1}{1.14}$, so that

$$t = \frac{2452}{1.14} - \frac{2452}{1.673} = 2151 - 1466 = 685'' = 11'25'';$$

the depths then being 0.5, 1, 1.5 feet; the corresponding intervals are $2'\ 31''$, $6'\ 4''$, $11'\ 5''$. If $h = 0$, it is evident that t becomes infinite, as $\dfrac{2452}{0} = $ infinity, and so also of any finite number in the numerator, arising from any other data. · If the depth sunk had been nearly equal to the whole charge at the commencement, as, suppose 2.4, so that $h = 2.8 - 2.4 = 0.4$, then $\dfrac{1}{\sqrt{0.4}} = \dfrac{1}{0.6324}$, and

$$t = \frac{2452}{0.6324} - \frac{2452}{1.673} = 3877 - 1466 = 2411'' = 40'\ 11''.$$

(XLII.) In question XIII., § 98, p. 126, taken from D'Aubuisson, the time of filling the lower part of a canal lock on the Canal du Midi, is calculated, i. e. up to the level of the centre of the sluices, placed in the

upper pair of gates; we can now, by the second case of
§ 95, calculate the time of filling up to the level of the
upper reach, from the centre of the sluice doors, which,
added to the 25″, as determined in XIII., will give the
total time. Substituting in the formula—

$$T = \frac{A}{mS \sqrt{2g}} \times \sqrt{H},$$

the several numerical values given at p. 91, we shall
have—

$$T = \frac{2 \times 3503.6}{0.548 \times 13^{\cdot}532 \times 8.024} \times \sqrt{6.3945},$$

that is

$$\frac{7007.2}{59.5} \times 2.53 \doteq 298 = 4' \, 58'',$$

to which adding 25″, we have 5′ 23″ as the total time of
filling a lock of such dimensions.

(XLIII.) The locks on the Montgomeryshire Canal
have a length of 81 and width of 7.75 feet; and at one,
named the Upper Belun Lock, the lift or rise was 7 ft.
A pipe leads the water from the upper level, and discharges below the surface of the lower level in the lock-
chamber, the diameter of which is 2 feet. As the mouth
of this pipe is a square, 2 feet in the side, gradually altered into a circular pipe, 2 ft. in diameter, we may take
$m = 1$, a result which is justified by comparing the observed time of filling this lock with that calculated by
the formula—

$$T = \frac{A}{mS \sqrt{2g}} \times \sqrt{H},$$

when m is put equal to unity, for

$$\frac{2 \times 81 \times 7.75}{1 \times 2^2 \times 0.7854 \times 8.024} \times 2.645 = 132'' = 2' \, 12'',$$

the observed time being 2′ 10″.

o

CHAPTER III.

FLOW OF WATER THROUGH PIPES, ARTIFICIAL CHANNELS, AND RIVERS.

110. GRAVITY is the sole force that acts upon a mass of water left to itself in a bed of any form; it produces all the motion which takes place,—the inclination of the surface of the water in the channel is the immediate cause of motion, being that which enables gravity to act: and thus the measure of this force is in feet per second, $g \times \sin i$,—in which g represents the measure of the force of gravity at the earth's surface, being the rate of motion at which a body is moving at the end of one second when falling freely in vacuo, or 32.1908 ft.; and i is the number of degrees, &c., of inclination of the surface of the water in the channel to the horizon; and $\sin i$ the ratio of the height fallen in any length to that length, or the fraction, $\dfrac{\text{height}}{\text{length}}$

Thus, if in one mile the surface was lowered 12 ft., we should have $\sin i = \dfrac{12}{5280}$, or $\dfrac{1}{440}$, and the constant dynamic force producing motion is measured by

$$g \sin i = 32.1908 \times \frac{1}{440} = 0.07316 \text{ ft. per sec.}$$

The angle of inclination being that which has the natural sine 0.0022727, or 0° 7′.45″. If, then, water flowing in a channel or pipe, and subject to this constant accelerating force, meet with no resistance, it will descend with an increasing velocity which would never be found uniform.

But observation and experience show that in open channels and pipes, even those of very great inclination, the rate of motion very soon becomes uniform. Bossut made the following experiment to prove this truth directly:—Having constructed a canal in wood, 650 ft. long, with a slope of 1 in 10, and marked off equal spaces of 108 ft. each, it was found that the water traversed each space, except the first, in equal times. There must then exist a retarding force, which destroys at each instant the effect of the accelerating force, and which, when the velocity has become uniform, is necessarily equal to it.

But in pipes, channels, &c., there can be no retarding force but that which arises from the resistance of the sides or bed : and of its existence we cannot doubt, for the simple experiment of the measurement of the discharge through a tube in a certain time, and again when the tube has been lengthened—all else remaining the same—proves that the time required to yield a certain volume of water has been increased also ; and this can only arise from the fact that the tube, or other channel, by reason of its increased length, offered a greater resistance to the velocity. The *surface* thus opposed motion.

To these retarding forces the name of Friction has been applied : though, from the difference between the laws of friction of water flowing over its resisting bed, and the friction of solid bodies sliding upon each other, we must look upon it as the application of an old word in a new sense, in preference to adding a new term to express this peculiar resistance. It may be useful to state here briefly the laws of friction of solid bodies, with the view of showing this contrariety.

111. *First Law.*—Experiment has shown that the friction or resistance to motion of bodies, sliding upon their surfaces of contact, is directly proportional to the force or weight pressing the two surfaces together, and differs only with the nature of the sliding surfaces, as wood, brass, iron, &c.

Second Law.—The amount of friction is independent of the extent of the surface pressed, provided the whole amount of the pressure remains the same, and that the substance of the surface pressed is the same.

Third Law.—The friction of a body, when in a state of continuous motion, bears a constant ratio to the pressure upon it, which is the same, whatever may be the velocity of the motion,—it is, in other words, independent of the velocity. Thus the first only of these laws can be expressed algebraically.

112. In the case of fluids, it has been shown that the resistance to motion, which we observe, and which has been called friction also, is, on the contrary—

First Law.—Independent of the pressure, that is, that the resistance to motion in a pipe with a head or pressure of, suppose, 100 ft., is the same as if the head were but 50 ft., or any other height, the velocity being the same, Dubuat had proved this by experiments on the oscillation of water in syphons, which has been thus modified :—

Two vessels ABCD, *abcd* (Fig. 69), were connected by the bent pipe EFG *gfe*, which turned round in the short tubes E and *e*, without

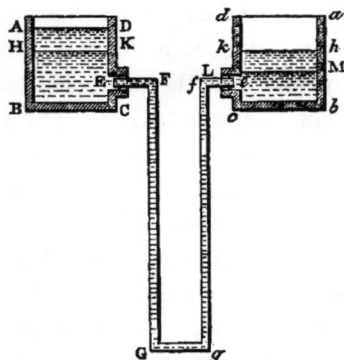

Fig. 69.

allowing any water to escape; the axis of these tubes

being in one right line. The vessels were about 10 inches deep, and the branches FG, *fg* of the syphon were about 5 feet long. They were then set on two tables of equal height, and (the hole *e* being stopped) the vessel ABCD, and the whole syphon, were filled with water, which was also poured into the vessel *abcd* till it stood at a certain height LM. The syphon was then turned into a horizontal position, and the plug drawn out of *e*, and the time carefully noted which the water employed in rising to the level HK*kh* in both vessels. The whole apparatus was now inclined so that the water ran back into ABCD. The syphon was now put in a vertical position, and the experiment repeated: no sensible or regular difference was observed in the time; yet in this experiment the pressure on the part G*g* of the syphon was more than six times greater than before. As it was thought that the friction on this small part (only 6 inches) was too small a portion of the whole resistance, various additional obstructions were put into this part of the syphon, and it was even lengthened to 9 feet; but still no remarkable difference was observed. It was even thought that the times were less when the syphon was vertical; nor has any variation ever been observed in the friction of water in these different positions when the surface was glass, lead, iron, wood, &c. (Principes d'Hydraulique, tome i., §§ 34 and 36, Dubuat.)

Second Law.—The resistance is, at any one velocity, proportional to the surface exposed to the action of the flowing water. In order to obtain an expression for this law, we may remark, in the first place, that in any channel or pipe the resistance arising from the surface is shared by all the particles in the volume of water flowing down, those nearest the sides being most retarded, and each in succession less and less influenced. This is

proved by the result of observations shown in the engraving, Fig. 70, which represents the transverse section
of a trapezoidal channel, with lines of equal velocity
plotted upon it, as given in the recent work of M. Darcy
and M. Bazin. The width of this experimental chan-

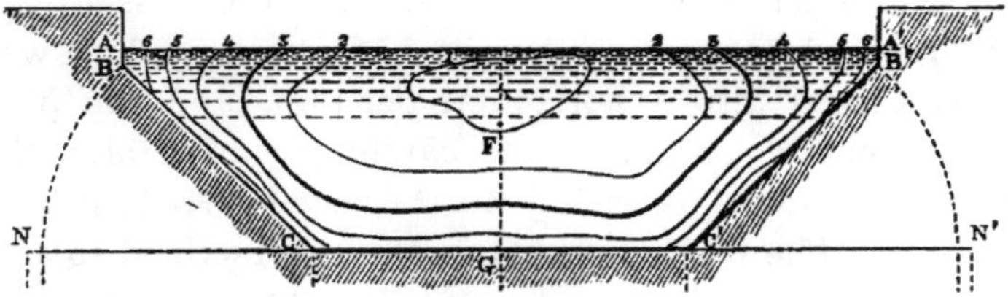

Fig. 70.

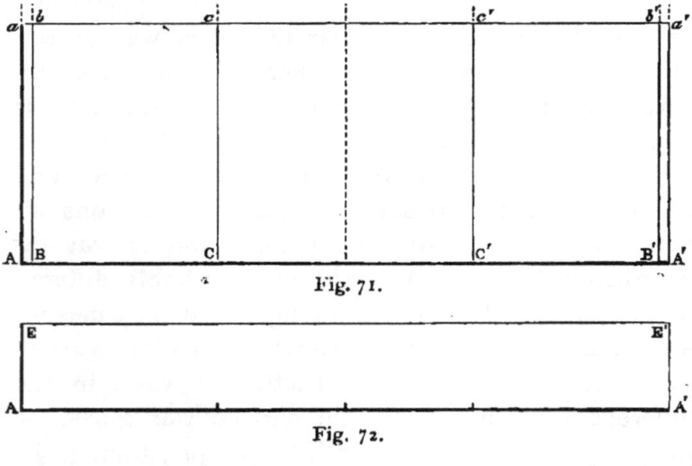

Fig. 71.

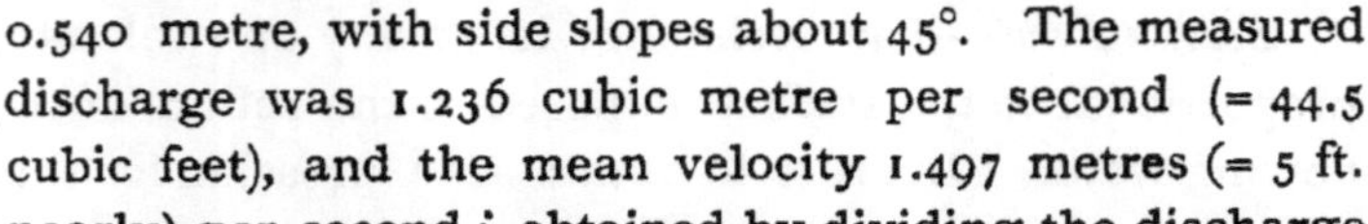

Fig. 72.

nel at the water surface was 2 metres, *qp*, and its depth
0.540 metre, with side slopes about 45°. The measured
discharge was 1.236 cubic metre per second (= 44.5
cubic feet), and the mean velocity 1.497 metres (= 5 ft.
nearly) per second ; obtained by dividing the discharge
by the area of the transverse section, which was equal
to very nearly 0.824 square metre. By improvements
on Pitot's tube (p. 162) this instrument was adapted by
them to the accurate measurement of the velocity in any

part of the transverse section, and from the observations thus taken the lines of equal velocity were plotted (by a method described further on). The darker line, No. 3, shows the points in the flowing water at which the mean velocity of 1.497 metres per second was found. The line, No. 1, which returns upon itself, shows continuously the points of highest velocity plotted ; No. 2 being also greater than the mean, while lines, Nos. 4, 5, and 6, show the successively decreasing velocities below the mean, the least being that nearest the surface of the sides and bottom. It would be easy to interpolate by hand any number of intermediate lines of equal velocity, and thus divide the whole mass of moving water into successive lamina, each suffering less resistance than the previous one as we proceed from the wetted surface of the bottom and sides inwards. The point of maximum velocity was situated on the central dotted line about one-third of the depth from the surface, and was equal to 1.82 metre per second. The greater, then, that *surface* is, the greater is the resistance. But the greater the *volume* upon which this retarding action of the surface has to act, the less reduced will be the velocity of the first, and therefore of each successive lamina : and thus we have the resistance directly proportional to the surface and inversely as the volume, i.e. proportional to $\frac{\text{area of sides and bottom}}{\text{volume of moving water}}$. Now let us suppose the channel, Fig. 70—which is identical in every section throughout its length, and having a uniform flow—to be cut by two parallel planes perpendicular to the axis of the stream ; and in the plan, Fig. 71, let $\overline{aa'}$ and $\overline{AA'}$ be the horizontal traces of these two planes, and let the base of the section, Fig. 70, be produced on each side until the produced part, CN and C'N', equal the sum of the sloping sides and short vertical portions, BA and B′A′. If, then, from the extremities N and N′ of this line perpendiculars be let fall on the traces, Fig. 71, the rectangle aAA′$a′$ so formed is evidently equal to the wetted surface of the channel between the two planes, that is, to the product of the distance between them, $\overline{a\mathrm{A}}$ and $\overline{\mathrm{A A}'}$, = $\overline{\mathrm{NN}'}$; also the volume of water between the same planes is equal to the product of $\overline{a\mathrm{A}}$ into the transverse section of the channel. Hence the ratio given above is equal to

$$\frac{\overline{a\mathrm{A}} \times \overline{\mathrm{NN}'}}{\overline{a\mathrm{A}} \times \text{transverse section}}.$$

Striking out from each the length aA of the channel common to both, we have the resistance directly proportional to the border or wetted perimeter, and inversely as the area of the transverse section perpendicular to the axis of the stream. If, then, we put C for the contour of the border, and S for the area of section, we have the resistance proportional to $\dfrac{C}{S}$.

Third Law.—The resistance is proportional to the square of the velocity nearly, the border being constant. For the *number of particles* drawn in one second from their adhesion to the sides of the channel or pipe is proportional to the number of feet per second with which the water is moving, that is, to the velocity. And the *force* with which they are drawn is also as the same number of feet per second, or the same velocity : and thus the passive resistance of the wetted border to the flow of the water is proportional to the product of the velocity into the velocity ; this part, then, of the expression for the resistance is represented by av^2, a being a constant, determined hereafter.

Experimenters have shown that this gives the resistance a very little too high, and that with velocities increased in the ratio 2, 3, 4, &c., it is not represented by $a \times 4$, $a \times 9$, $a \times 16$, &c., but more nearly by adding the simple power of the velocity, thus $a (v^2 + bv)$, the series of numbers $v^2 + v$ not increasing so fast as v^2.

Fourth Law.—In gases and elastic fluids we also have the friction proportional to the specific gravity or density.

In order to obtain from these laws a formula for the discharge of water through pipes and channels, we must make use of the well-known principle, that when any body is moving with a uniform velocity, the accelerating are necessarily equal to the retarding forces: for if the accelerating forces be supposed greater than the retarding, the velocity must increase; and if they should become less, then the velocity must, on the other hand, decrease. We must now, as in Chapters I. and II., find a general expression for the mean velocity, for this multiplied into the transverse area gives the discharge with a given inclination: and we can thus solve the questions that arise in practice, such as the requisite dimensions of pipe or channel to convey a given quantity of water, &c., &c.

Now in any pipe or channel, whose length is l, and whose height, from the surface of the supply to the point of discharge or extremity of l, is represented by h, we have the accelerating force expressed by $\dfrac{h}{l} \times g$, or sine of inclination of surface into gravity.

The retarding forces are, from the second and third laws above given, neglecting bv, proportional to

$$(1) \quad \cdot \quad \cdot \quad \cdot \quad \quad \cdot \quad \cdot \quad \cdot \quad \frac{C}{S} \times v^2,$$

and therefore we have

$$(2) \quad \ldots \ldots \quad g \times \frac{h}{l} = \frac{C}{S} \times a \times v^2.$$

Each side of this equation represents an equal number of feet per second. The left-hand being 32.1908 ft. per sec. reduced, by being multiplied by a fraction whose value depends on the inclination of the surface, that is $\frac{\text{height}}{\text{length}}$. And the right-hand side being the square of the number of feet per second with which, at a mean, the water is moving when the motion has become uniform, reduced by the constant multiplier a, and also by a quantity depending on the figure of the transverse section of the channel,—a being some constant quantity to be determined by experiment. If the formula be correct, all good experiments will give the same value for a, that quantity by which the right-hand side of equation (2) must be multiplied to produce the equality. We may, however, simplify the expression by dividing out by g, and thus we have—

$$(3) \quad \ldots \ldots \quad \frac{h}{l} = \frac{a}{g} \times \frac{C}{S} \times v^2,$$

and as g is constant, put $\frac{a}{g} = a'$, which must be constant if a be so; solving, then, for a', we have—

$$(4) \quad \ldots \ldots \quad \frac{h}{l} \times \frac{S}{C} \times \frac{1}{v^2} = a'.$$

Substituting the data of experiments in the left-hand side, and deducing v from $Q' \div TS = v$, we obtain a', and comparing different experiments, we find that it remains very nearly the same in all.

The celebrated Smeaton has given in his Reports (vol. ii., p. 297) a series of experiments on the velocity of water flowing through pipes under pressure. One of these had the following data :—Diameter of pipe, $4\frac{1}{2}$ inches, or 0.375 ft.; length, 14637 ft.; fall or head, 51.5 ft.; and $v = 1.815$ ft.   Hence

$$\frac{51.5}{14637} \times \frac{0.375^2 \times 0.7854}{0.375 \times 3.1415} \times \frac{1}{1.815^2} = \text{nearly, } 0.0001 = a'.$$

The quantity discharged is given by Smeaton in Scotch pints, which he states contain 103.4 cb. inches, and therefore, the number of cb. ft. in one pint is $\frac{103.4}{1728} = 0.05984$, and as 200 pints per minute were discharged, we have $Q' = 11.968$ cb. ft.   Hence as $\frac{Q'}{T \times S} = v$, we have $\frac{11.968}{60 \times 0.110247}$ $= 1.815$ ft. per second.

Mr. Provis has published in the " Transactions of Civil Engineers," vol. ii., p. 203, some experiments on the flow of water through pipes $1\frac{1}{2}$ inch $= 0.125$ ft. in diameter; of these, No. 4, with a length of 100 ft. delivered 2 cb. ft. *per minute*, with a head of 2.5 ft.   (It is presumed that the orifice of entry of the water was of the best form.)   Here the velocity will be (2 cb. ft. $\div$ 60 $\times 0.125^2 \times 0.7854 =$) 2.72 ft. per second: and hence the value of a' is found

$$\frac{2.5 \text{ ft.}}{100 \text{ ft.}} \times \frac{0.125^2 \times 0.7854}{0.125 \times 3.1415} \times \frac{1}{2.72^2} = \text{very nearly, } 0.0001;$$

and as from these and many other experiments $a' = 0.0001$, we have $\frac{1}{a'} = 10,000$.

Substituting, then, this value of a', and solving the equation (4) for v, we have the following expression for the mean velocity

$$v = \sqrt{\frac{h}{l} \times \frac{S}{C} \times 10000};$$

or, taking the root of the factor 10,000, and placing it outside,

$$(5) \quad \ldots \quad v = 100 \sqrt{\frac{h}{l} \times \frac{S}{C}}.$$

From this expression it is evident that many geometrical questions arise in designing the best form of channel, whether circular, rectangular, or trapezoidal, to convey given quantities of water: a given area having, with the same condition as to ratio of slopes, a great number of different borders, and one a minimum, and, *vice versâ*, a given border, having a number of different sectional areas, and one a maximum.

The quantity $\frac{S}{C}$ has been called the hydraulic mean depth or mean radius; it is, in every form and section of channel, represented by a line AE, Fig. 72; the rectangle under which, and the border ABCC′B′A′, (extended into one right line, AA′ = NN′), is equal to the area of the section; the greater it is, the less the relative resistance of the surface to the volume of water passing over it. It is important, therefore, to have a clear idea of the influence of the figure of the transverse section of the channel upon the magnitude of this quantity, on which, other things being the same, the mean velocity depends, being directly proportional to its square root. As a simple form, let us take a channel whose transverse section is a rectangle, and, first, suppose the *border* to be constant, secondly, the *area*.

Now when the border is constant, it is evident that there are two extreme positions of the figure: one, when the depth becomes zero, in which case the bottom width must equal the constant border, as suppose 200 ft., or yards or metres, and coincide with the line of water sur-

face. This is shown in Fig. 73, in which the line ACB
is the level of the surface of the water of the several
transverse sections, and CD the vertical central line
with reference to which they are all symmetrically arranged. The line ACB represents 200 ft., and is, as
has been stated, the limiting figure when the depth

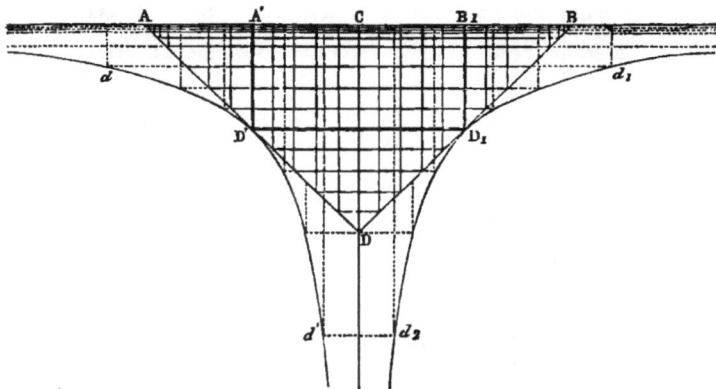

Fig. 73.

becomes zero. The other extreme position of the figure
of the rectangular section having a constant wetted border is when the bottom width becomes zero, and then
the depth must be equal to half the border, or 100 ft.,
and coincide with the vertical central line C D.

Now if we take any other rectangle, as $A'D'D_1B_1$,
having the same border, and its vertical central line
coinciding with CD, the water-surface line also coinciding with ACB, we shall have $AA' = A'D'$, and BB_1
$= B_1D_1$, and therefore the points D' and D_1 are in the
right lines joining the point D and the points A and B :
and thus all possible rectangles having this constant
border may be inscribed in the isosceles right-angled
triangle ADB.

The following Table gives the dimensions of those

transverse sections which are drawn in Fig. 73 within
the triangle ADB :—

TABLE *showing the Value of the Hydraulic Mean Depth,
Area, &c., the Channel having a Rectangular Transverse
Section, the Border being constant.*

Depth in Feet.	Bottom Width in Feet.	Area in square Feet.	Hydraulic Mean Depth or Area ÷ 200.	Square Roots of Hydraulic Mean Depths.
0	200	0	0.00	0.00
1	198	198	0.99	0.99
2	196	392	1.96	1.40
3	194	582	2.91	1.84
5	190	950	4.75	2.18
10	180	1800	9.00	3.00
20	160	3200	16.00	4.00
25	150	3750	18.75	4.33
30	140	4200	21.00	4.58
40	120	4800	24.00	4.90
50	100	5000	25.00	5.00
60	80	4800	24.00	4.90
70	60	4200	21.00	4.58
75	50	3750	18.75	4.33
80	40	3200	16.00	4.00
90	20	1800	9.00	3.00
95	10	950	4.75	2.18
97	6	582	2.91	1.84
98	4	392	1.96	1.40
99	2	198	0.99	0.99
100	0	0	0.00	0.00

It will be perceived at once that the area and hydraulic
mean depth increase progressively up to that figure in
which the depth is equal to half the bottom width, the
rectangle being half of the square whose side is the
bottom width: and although the ratio of the hydraulic
mean depth to the depth is continually decreasing, yet
the former quantity increases until this ratio has become ½, after which point it diminishes, having the same

value at depths equidistant from the maximum. The inner curve CND, Fig. 74, gives a diagram representation of the results of the Table. The depths are severally plotted on the line CD from C as zero, (about $\frac{1}{3}$ larger than in Fig. 73 for the sake of clearness), and the hydraulic mean depths as vertical ordinates at the points in CD, corresponding to the depths. Joining

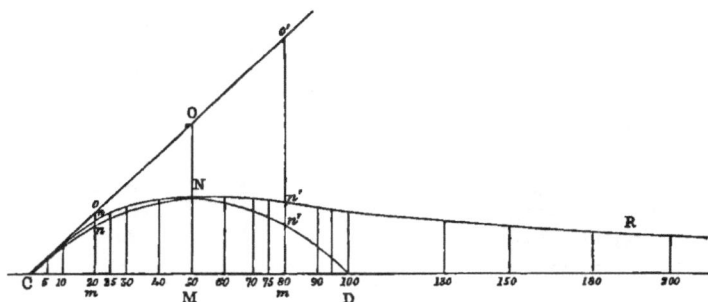

Fig. 74.

the termination of these ordinates, we have the curve line CND, and by it we can obtain the hydraulic mean depth due to any particular depth by drawing a vertical line up to the curve at the point.

If, instead of a constant border, we assume the rectangular transverse section to have a constant area, and for readier comparison with the foregoing take 5000 sq. ft., that area, namely, which was a maximum with the border of 200 ft., we now find that, with a depth indefinitely small, the bottom width must be indefinitely great ; and when we assume a bottom width indefinitely small, then the depth must be indefinitely great, as the product in each case is a given quantity : and rectangles, intermediate between these extremes, being placed, as before, with respect to the lines AB and CD, Fig. 73, we shall find the points corresponding to D′

and D_1 lie in a curve, well known as the hyperbola,
and represented in Fig. 73 by d, D', d'', and d_1, D_1, d_2;
the points D' and D_1 being common to the rectangle of
constant border and constant area.

TABLE *showing the Value of the Hydraulic Mean Depths,
&c., &c., the Channel having a rectangular Transverse
Section; Area constant.*

Depth.	Bottom Width.	Hydraulic Mean Depth.	Square Roots of Hydraulic Mean Depth.	Border.
0	∞	0	0	∞
1	5000	0.9994	0.9997	5002
2	2500	1.997	1.413	2504
3	1666.6	2.989	1.726	1672.6
5	1000	4.95	2.225	1010
10	500	9.604	3.878	520
20	250	17.24	4.152	290
25	200	20	4.472	250
30	166.6	22.065	4.700	226.6
40	125	24.39	4.939	205
50	100	25	5	200
60	83.33	24.564	4.956	203.33
70	71.4	23.65	4.863	211.4
75	66.6	23.084	4.800	216.6
80	62.5	22.47	4.74	222.5
90	55.5	21.23	4.607	235.5
95	52.6	20.61	4.54	242.6
98	51.02	20.24	4.50	247.02
99	50.505	20.12	4.48	248.505
100	50	20	4.472	250
1000	5	2.494	1.58	2005

All possible rectangles having this constant area
may be inscribed in the space formed by the two
branches of the curve and the right line A, C, B, produced each way indefinitely. The Table gives the dimensions of some of those drawn in Fig. 73 within this
space, the sides having dotted lines.

An inspection of this Table shows that when the area of the rectangle is constant, the hydraulic mean depth increases with the increase of the depth, being at corresponding depths somewhat greater than in the former Table, except at the maximum value, which is, by construction, the same in each : and at this point, as before, the hydraulic mean depth is half the depth.   The outer curve CNR, Fig 74, gives a diagram of the results, the hydraulic mean depths being plotted, as before, at the depths of channel, from which they were calculated. The curve, commencing at zero at the same point C, also passes through a common point N, but from this it diverges, and soon becomes convex to the line CD produced, which it never can reach, the hydraulic mean depth having always a finite value, in the case of a constant area, as long as the depth is finite.   In every part this last curve is exterior to that representing the results of a rectangle with constant border, coinciding only at the points C and N.

If we produce MN, so that MN = NO, and draw the line CO, producing it indefinitely, then the ordinates, as *mn* or *m'n'*, being produced to cut this line in *o* and *o'*, we have the ratio of the hydraulic mean depth to the depth at each point, taking *n* and *n'* either on the inner or outer curve.

It will be proved generally in § 115 that the best form of channel, whether the transverse section be rectangular, polygonal, or circular, is when half the depth of the water at the centre line is equal to the hydraulic mean depth ; a proposition which has appeared in the above Tables for the particular numbers chosen, which are mainly intended to illustrate the importance of the figure of the transverse section of a rectangular channel in regard to the velocity and discharge.

Let us suppose that, with the same inclination, we

P

had two rectangular channels of equal transverse area;
but in one the depth and bottom width were 5 ft. and
1000 ft. respectively, and in the other 50 ft. and 100 ft.,
or numbers in those ratios; then, from the second Table,
we perceive that the square roots of the hydraulic mean
depths are as 2.214 to 5, and therefore the mean velocity,
which is proportional to this quantity, is more than
double, and the volume of water flowing down, which
is the product of the mean velocity into the transverse
area, also more than twice as great. When the border
is constant, the comparison gives results still wider.
Thus, if from the first Table we take rectangular channels, whose depth and bottom width are 10 and 180,
and again 50 and 100 respectively, or any numbers in
those ratios, we find the square roots of the hydraulic
mean depths are as 3 and 5, and, multiplying each into
the area, we have the volumes of water carried down as
5400 to 25000, the inclinations being supposed the same
in both channels.

We may also gather from these Tables, that in wide
rivers and channels in which the depth is small compared with the width, the mean velocity is very nearly
proportional to the square root of the depth, for in such
cases the hydraulic mean depth is nearly equal to the
depth, as in the upper lines of each Table. It is also
evident that on each side about the maximum value the
mean velocity does not diminish very rapidly; thus, in
the second Table, the rectangles, 40 by 125, and 60 by
83.3, differ but very little in their mean velocity, and
therefore in volume discharged from that of 50 by 100.
In the first Table, in like manner, the mean velocities,
for depths intermediate between 40 and 60, being nearly
identical with the maximum at 50, the volumes discharged will only vary from the maximum discharge in
proportion to the areas, and thus, in practice, the advan-

tages of the best form of channel may, in a great degree, be obtained by others chosen within a considerable range on each side of it.

In the case of tubes having a uniform circular section. $\dfrac{S}{C} = \dfrac{d^2 \times 0.7854}{d \times 3.1415} = \dfrac{d}{4}$, the formula (5) becomes then

in the case of pipes flowing full—

(6) . . $v = 100 \sqrt{\dfrac{h}{l} \times \dfrac{d}{4}} = 50 \sqrt{\dfrac{h}{l}} \times d$ ft. per sec.

We have seen, in speaking of the second law of friction, that each successive couche or lamina, into which we may suppose the fluid in motion to be divided, is less and less retarded from the border towards the centre of the section : the highest velocity being consequently near the centre and in open channels a little below the surface. The volume of water which traverses the section of which we speak, in one second, is due to these different velocities; and the velocity, the expression for which has now been determined, is that one of these various velocities with which, if the whole section moved as one solid mass, the discharge would be the same : it is then the mean velocity, and is found in any actual experiment by dividing the volume discharged in one second by the section, as has been done in the two experiments used for the determining the value of a'.

In order, then, to determine the discharge by any channel or pipe, for which we have deduced the value of v from the given inclination and hydraulic mean depth, we multiply the expressions (5) or (6) by the area. Thus from (5) we have

(7) . . . . $Q = S \times 100 \sqrt{\dfrac{h}{l} \times \dfrac{S}{C}}$;

or if we put H_y for $\dfrac{S}{C}$, the hydraulic mean depth,

$$Q = S \times 100 \sqrt{\frac{h}{l}} \times H_y \text{ cb. ft. per sec.}$$

And again, from (6) we have for pipes running full,

$$Q = 0.7854 \, d^2 \times 50 \sqrt{\frac{h}{l}} \times d,$$

or

$$(8) \quad . \quad . \quad . \quad Q = 39.27 \sqrt{\frac{h}{l}} \times d^5 \text{ cb. ft. per sec.}$$

In many works and reports the discharge is spoken of per minute, instead of per second: and for this unit of time we have $60 \times 39.27 = 2356.2$ as the factor outside; hence—

$$(8a) \quad . \quad . \quad . \quad Q' = 2356 \sqrt{\frac{h}{l}} \times d^5 \text{ cb. ft. per minute,}$$

which may be written thus,

$$(8b) \quad . \quad . \quad . \quad . \quad Q' = 2356 \times \frac{\sqrt{d^5}}{\sqrt{\frac{l}{h}}},$$

being the formula used by Beardmore in calculating Table 5, in his work.

If, as is not unusual, the diameter of the pipe be given in inches, which call d_1, the above equation becomes—

$$(8c) \quad . \quad . \quad . \quad . \quad Q' = 4.72 \sqrt{\frac{d_1^5 h}{l}} \text{ cb. ft. per minute,}$$

for this change in the units of the diameter is equivalent to multiplying the right-hand side of the equation by $\sqrt{12^5}$, or $\sqrt{248832} = 498.83$. In order, therefore, that Q' remain unaltered, we must divide the factor 2356 by this, and consequently, $2356 \div 498.83 = 4.72$, as above.

When the length of the pipes is given in yards, which call l_1, as is sometimes done in practice, we have the right-hand side of the last equation multiplied by $\sqrt{3} = 1.732$, by which, in order that Q' may remain the same, we must divide the numerical coefficient 4.72, which therefore becomes 2.725, and—

(8d) . . . . $Q' = 2.725 \sqrt{\dfrac{d_1^5 h}{l_1}}$ cb. ft. per minute.

And if in gallons per minute, which call $Q'' = Q' \times 6.25$, then both sides must be multiplied by 6.25, and we have—

(8e) . . . . $Q'' = 17.03 \sqrt{\dfrac{d_1^5 h}{l_1}}$ gallons per minute.

Again, if we find (though it cannot be said to be very common in practice) that the discharge is expressed in gallons per hour, we have, making $G = Q'' \times 60$, and multiplying by 60, we find, $60 \times 17.03 = 1021.8$—

(8f) . . . . $G = 1021.8 \sqrt{\dfrac{d_1^5 h}{l_1}}$,

(8g) . . . or $G = \sqrt{\dfrac{(16 d_1^5) h}{l_1}}$, nearly, as $1021.8^2 = 16^5$.

In some works we find the above stated as ($15 d_1^5$) which gives a less result.

An approximate practical rule of very easy application can be derived from equation (8f), by multiplying by 1000 and adding 2 per cent.

All these expressions from (8) have reference to pipes flowing full under pressure.

Other formulæ for the mean velocity, generally expressed in words, are in use amongst engineers, which are derived from (5) and those above given, namely, that the mean velocity of water in any pipe or channel that.has attained a uniform velocity is nine-tenths of the square root of the product of twice the fall per mile into the hydraulic mean depth ; or sometimes thus expressed, 0.92 into a mean proportional between twice the fall per mile and the hydraulic mean depth.

These, which would not be given in words but to obviate any disadvantage arising from the student meeting with them so expressed, are consequences of equation (5) ; for the numerator and denominator of the fraction $\dfrac{h}{l}$ may be replaced by any numbers having the same ratio. If, then, we make $l = 5280$, i. e. the number of feet in a mile: the numerator, which we may call f, expresses the fall per mile thus, $\dfrac{h}{l} = \dfrac{f}{5280}$; and from (5) we have, therefore,

$$(9) \quad v = 100 \times \sqrt{\frac{f}{5280} \times H_y} = \frac{100}{72.66} \times \sqrt{fH_y} = 1.38 \sqrt{fH_y};$$

and as $1.38 = 0.92 \sqrt{2}$, we have, by substituting this value in (9),

$$(10) \quad \ldots \ldots \quad v = 0.92 \sqrt{2fH_y} \text{ ft. per sec.}$$

This is sometimes written in the nearly identical form—

$$(10a) \quad \ldots \ldots \quad v = \frac{10}{11} \sqrt{2fH_y} \text{ ft. per sec.}$$

And if the velocity be expressed in feet per minute, we have, since $0.92 \times 60 = 55.2$—

$$(10b) \quad \ldots \ldots \quad v' = 55 \sqrt{2fH_y} \text{ ft. per minute,}$$

the decimal being neglected.

Again in Dr. Young's " Natural Philosophy " we find this rule :—" The square of the velocity in any measures per sec. is equal to the product of the fall in 2800 yards into the hydraulic mean depth, all in the same units. For if f_1 be the fall in 2800 yards, and f that in one mile, as above, then, since $\dfrac{2800}{1760} = 1.59$, we have $f_1 = 1.59 \times f$ also $\sqrt{1.59} = 1.26$. If, now, we take the coefficient in equation (10) as being 0.9 instead of 0.92, we may, since $0.9\sqrt{2} = 1.27$ express it thus: $v = 1.27\sqrt{fH_y}$. In this changing f the fall per mile into f_1 the fall in 2800 yards, we have $v = 1.27\sqrt{\dfrac{f_1}{1.59} \times H_y}$, or $v = \dfrac{1.27}{1.26}\sqrt{f_1 H_y}$. The quotient of these numbers is so nearly unity that we may assume the equation—

(10*c*) . . . . . $v = \sqrt{f_1 H_y}$, which being squared

gives the above rule, and shows that this eminent author used the same coefficient of resistance as has been deduced in equation (5), p. 204.

113. From the formula (8) to (8*f*) for the discharge of pipes running full under pressure, we can, being given any two of the three quantities Q, the inclination $\dfrac{h}{l}$, or d, determine the other. Let it be required to find the diameter of the pipe, which, with a given inclination, shall convey a given quantity of water. Dividing equation (8) by 39.27, and squaring both sides, we have—

(11) . . . . . . $\left(\dfrac{Q}{39.27}\right)^2 = \dfrac{h}{l} \times d^5$,

and dividing by $\dfrac{h}{l}$, or multiplying both sides by $\dfrac{l}{h}$, and extracting the fifth root,

(12) . . . . . $\sqrt[5]{\left(\dfrac{Q}{39.27}\right)^2 \times \dfrac{l}{h}} = d.$ ft.

The requisite inclination is found from (11) by dividing both sides by d^5,

$$(13) \ldots \ldots \left(\frac{Q}{39.27}\right)^2 \times \frac{1}{d^5} = \frac{h}{l};$$

and if we multiply both sides by l, we obtain h: so that if the length the water has to be conveyed be also amongst the data, we obtain the head or pressure necessary to force the given quantity along a pipe of known length and diameter—

$$(14) \ldots \ldots \left(\frac{Q}{39.27}\right)^2 \times \frac{l}{d^5} = h \text{ ft.}$$

We cannot, however, fully determine the figure of a rectangular or trapezoidal channel from (7); solving it for $\frac{S^2}{C}$ we have

$$(15) \ldots \ldots \left(\frac{Q}{100}\right)^2 \times \frac{l}{h} = \frac{S^2}{C}.$$

In this we require, in addition, to be given either S or C, and also the ratio of the slopes of the sides if it be a trapezium; moreover, S and C are so related that, with given slopes, there is a maximum value of S to every given value of C; if S exceed this maximum, the solution is impossible.

114. It is found in practice that certain soils, in every excavation for whatever purpose, require a rate of slope in the sides adapted to the degree of cohesion of the ground, to obviate the danger of slips, which occur when they are too steep: this slope of the banks is, therefore, always found amongst the requisite data in the designing of channels being trapeziums in transverse section.

In order that the side slopes of channels, intended to be permanent, may stand without any masonry or dry stone pitching, they should have a slope between the rates of

1½ horizontal to 1 vertical, and 2 to 1 : being made flatter according as the soil has less tenacity. In some cases even 2½ to 1 has been adopted; the half regular hexagon has slopes of 0.58 to 1 ; in channels for temporary use we may have 1 to 1.

And so also must the velocity be given ; and, for the same reason, some kinds of earth being worn away, and the form of channel destroyed, by a rate which carries down the particles of the soil through which it is excavated, a velocity must therefore be assigned within this rate of motion. It has been determined experimentally for many kinds of earth.

The effect of the velocity of the water, in carrying down the particles of the ground through which the channel is excavated, depends jointly upon their tenacity and size. As to the size, we know that the cubical quantities or weights of any similar bodies decrease faster than their superficial areas ; and the pressure or force urging a body down stream being, *ceteris paribus,* proportional to the surface, is relatively greater the less the volume; the smaller the particles, therefore, the less is the velocity required to move them. Mr. Beardmore* in Table 3 gives the following statement of the limit of *bottom* velocities in different materials in feet per minute :—

30 ft. will not disturb clay with sand and stones.
40 ,, will move along coarse sand.
60 ,,   ,,   ,,   fine gravel, size of peas.
120 ,,   ,,   ,,   rounded pebbles, 1 in. diameter.
180 ,,   ,,   ,,   angular stones, about 1¾ in. do.

The beds of rivers, protected by aquatic plants, however, bear higher velocities than this Table would assign.

* Hydraulic Tables, by N. Beardmore.

Such being the natural limitations in the choice of any particular rectangle or trapezium, the engineer must proceed to determine the figure of the transverse area without violating the conditions they impose.

115. When it is desired to convey the greatest possible quantity of water in an open channel with a given area of transverse section, then the volume discharged being directly proportional to the area, and inversely as the wetted border, we must select the figure which for a given area has the least border, and for a given border has the greatest area.

Geometry informs us that the circle has this property: the semicircle, and therefore the semicircular channel, has the same property; the ratio between the area of the semicircle and semi-circumference being the same as that between the circle and the entire circumference. Then follow the regular demi-polygons, with less and less advantage as the number of their sides is less ; and among the more practible forms are the demi-hexagon, and finally the half-square.

As the transverse sections of artificial open channels are, when without masonry, trapezoidal, the question as to the form of greatest discharge is reduced to taking among all the trapeziums with sides of a determinate slope, that which gives the greatest section for a given wetted border ; or, in other words, which has the greatest hydraulic mean depth ; and every different area and ratio of slopes has its particular maximum trapezium.

Let p be the depth of the trapezium BF (Fig. 75),

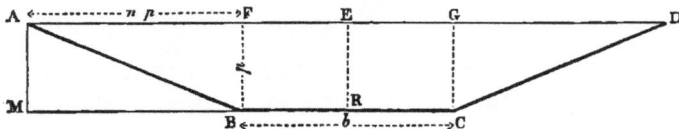

Fig. 75.

and b the bottom width BC, and $n : 1$ the ratio of the

slopes, or AF : FB; then the general values of S and C are

$$(16) \quad . \quad . \quad . \quad . \quad S = (b + np) \times p = bp + np^2,$$

and

$$(17) \quad . \quad . \quad . \quad . \quad C = b + 2p \sqrt{n^2 + 1}.$$

Since, then, S in the expression $\dfrac{S}{C}$, with slopes of $n : 1$, is a maximum, its differential will be zero, and we have

$$(18) \quad . \quad . \quad . \quad . \quad pdb + bdp + 2npdp = 0;$$

and as the border is constant, its general value being differentiated, gives $db + 2dp \sqrt{n^2 + 1} = 0$. Hence $db = - 2dp \sqrt{n^2 + 1}$; this being substituted in (18), gives $b = 2p \left(\sqrt{n^2 + 1} - n \right)$; with which value of b we have

$$(19) \quad . \quad . \quad . \quad \frac{S}{C} = \frac{p^2 \left(2\sqrt{n^2 + 1} - n \right)}{2p \left(2\sqrt{n^2 + 1} - n \right)} = \frac{p}{2}.$$

Therefore in all trapezoidal channels of the best form, with certain given slopes and area, the hydraulic mean depth is half the depth of the water: and hence we derive a construction for the cross section of a maximum discharging channel; remarking that as $\dfrac{S}{C} = \dfrac{p}{2}$, we have

$S = C \times \dfrac{p}{2}$. Let the trapezium ABCD (Fig. 76) be the

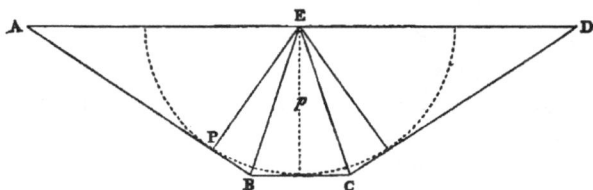

Fig. 76.

channel sought; from the middle point E of the top

width draw lines EB and EC dividing the figure into three triangles, of which AEB and CED are identical; let EP be the perpendicular from E upon AB; then

$$\overline{AB + BC + CD} \times \frac{p}{2} = \overline{AB + CD} \times \frac{EP}{2} + BC \times \frac{p}{2};$$

and therefore $\frac{p}{2} = \frac{EP}{2}$. Hence from E as centre, and with p as radius describing a circle, it will touch the two sides AB and CD. If, therefore, conversely, we describe a circle (Fig. 77) with any radius, and draw a tangent, parallel to a horizontal diameter, produced on each side indefinitely, and then between these lines draw tangents having the given inclination, we obtain a figure similar to that re-

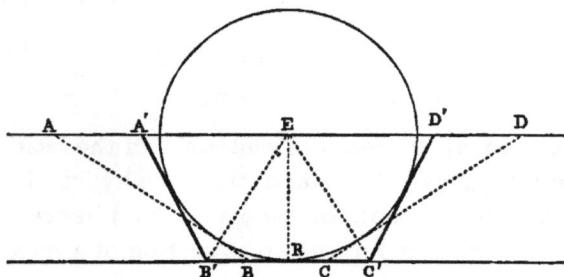

Fig. 77.

quired, from which, by proportion, we find the transverse section of the channel sought: a construction given by Mr. Neville in his Hydraulic Tables.

Other properties of the trapezium of greatest discharge, Figs. 76 and 77, are, First, that the line of surface of water AD or A'D', Fig. 77, is equal to the sum of the slopes AB and CD, or A'B' and C'D', and consequently the wetted border is equal to the sum of the top and bottom widths, or the mean breadth equal half the border. Secondly, the triangle BEC, Fig. 76, is similar to the triangles EAB and EDC; the vertical angle BEC being equal to the angle of inclination of the sides to the horizon. Thirdly, the angle between the perpendiculars from E upon the sides AB and CD is double the angle of inclination of the sides, and the angle PEB half of the same, that is, of the angle BEC. This gives another construction when p and the angle of inclination of sides are given. On a vertical line lay off p, and from the upper point E (Fig. 76), on each side, lay off the angle of inclination, bisect each of them by EB and EC, and through the lower point of p draw a perpendicular to intersect EB and EC, which gives the base BC, then from B draw BA perpendicular to EP, to intersect the horizontal line through E at A, and in like manner on the opposite side, giving the required trapezium ABCD. From the second property we obtain an expression for the area of the trapezium of greatest discharge in terms of the depth and angle of inclination β of the side slopes with the horizon, for the area of the triangle BEC is equal to $p \times p \tan \frac{1}{2}\beta$, as half $\overline{BC}$ is the tangent of half the vertical angle to radius p; also the sum of the areas of the triangles EAB and EDC is equal to $\overline{EP} \times \overline{AB}$, but EP is equal to p, and $\overline{AB}$ is the cosecant of β to radius p, as is evident if from B, Fig. 75, we draw the perpendicular BF, the angle ABF being the complement of BAF, that is β; thus the area of the trapezium is

$$p^2 (\tan \tfrac{1}{2}\beta + \text{cosec.} \, \beta).$$

We may deduce from the Table that very large channels formed in any kind of earth cannot be designed so as to be of the best discharging form, as the depth of excavation would be too great ; the ratio of the depth to the mean width must rather resemble that observed in large rivers.

TABLE *giving the Values of* tan $\frac{1}{2}\beta$ + cosec β, *and the Top and Bottom Widths, in Trapeziums of best Form and ordinary Slopes.*

Slopes.	Angle β.	Tan$\frac{1}{2}\beta$ + cosec β.	Top Width,	Bottom Width.
0 to 1	90°	2.000	$p \times 2.000$	$p \times 2.000$
$\frac{1}{4}$ "	75 58′	1.812	2.062	1.562
$\frac{1}{2}$ "	63 26	1.736	2.236	1.236
$\frac{4}{7}$ "	60	1.732	2.309	1.155
$\frac{3}{4}$ "	53 8	1.750	2.500	1.000
1 to 1	45	1.828	2.828	0.828
$1\frac{1}{4}$ "	38 39	1.952	3.202	0.702
$1\frac{1}{3}$ "	36 53	2.000	3.333	0.666
$1\frac{1}{2}$ "	33 41	2.106	3.606	0.606
$1\frac{3}{4}$ "	29 44	2.282	4.032	0.532
2 to 1	26 34	2.472	4.472	0.472
$2\frac{1}{4}$ "	23 58	2.674	4.924	0.424
$2\frac{1}{2}$ "	21 48	2.885	5.385	0.385
$2\frac{3}{4}$ "	19 58	3.104	5.854	0.354
3 to 1	18 26	3.325	6.325	0.325

The several numbers in the third column express the areas of the trapeziums of best form to the depth unity, and also the mean widths: multiplied by p^2 they give the area, and by p the mean width for the depth p. The numbers under the fourth and fifth columns are the top and bottom widths to the depth unity, and for any depth p give the top and bottom width respectively, by being multiplied into p; they are obtained by adding, for the fourth column, the numbers in the first and third, and for the fifth column, by subtracting the same, for in every trapezium the top width is equal to the mean width, plus the depth into the ratio of the slopes, and the bottom width equal to the mean, minus the depth into the ratio

of the side slopes, and with a depth unity the value of
these lines is derived by the addition and subtraction
above mentioned.

In all the different trapeziums of best discharge

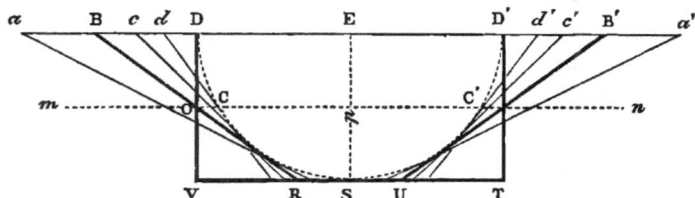

Fig. 78.

formed, as shown in Fig. 78, by drawing tangents to
the same semicircle having the radius p, the hydraulic
mean depths are evidently the same, namely, $\frac{1}{2}p$, whatever the side slopes may be, and, therefore, with the
same inclination of the bed of the channel, they all have
the same velocity, and consequently the discharge,
which will then be as the area, is proportional to the
mean width, the depth p being constant.

If through the middle point of p we draw the indefinite line mn parallel to aa', then the areas of all these
figures formed by the several tangents to the semicircle
will be proportional to the length of this line cut off by
the tangents or side slopes; this consideration serves to
explain what may be observed in the last Table, namely,
that the numbers in the third column decrease from the
first number, corresponding to 0 to 1, down to a certain
point, and then, rising to the first value at the slope of
$1\frac{1}{2}$ to 1, afterwards increase continuously. In the woodcut the vertical tangents at D and D' form the rectangle
of best discharge with a given area, DV × VT, and the
length of mn cut off, which is the measure of the area,
is equal to twice the depth p; now every other tangent between that from D and that from B, which passes

through the middle point O of DV (and, therefore, forms
a trepezium BRUB', having the same mean width as the
rectangle), cuts off a smaller part of the line mn, and
so is of less area than the rectangle, the minimum
being the trapezium formed by the tangents dC and d'C',
which touch at the points C and C' in which the line
mn cuts the semicircle, and having therefore the inclination of 60° with the horizon, the mean width, to the
depth unity, being 1.732, and the trapezium a half regular hexagon; but the tangents cutting mn beyond the
point O form trapeziums of a continually increasing
mean width, and, therefore, increase in area in the same
proportion. If through the point O we draw any inclined lines whatever, the areas of the figures so formed
are all equal to that of the rectangle, but that particular
line which, drawn through O, also touches the semicircle, forms a trapezium, whose wetted border, as
well as area, is equal to that of the rectangle, for the
slope BR is equal to the half top width BE, which is
equal to the radius DE (that is p or DV), together with
BD, which is equal to VR, and, therefore, BR is equal
to DV + VR, and consequently $\overline{BRUB'} = \overline{DVTD'}$.

Also the slope of BR is $1\frac{1}{3}$ to 1, or, which is the same
thing, VR = $1\frac{1}{3}$ × OV; for the sum of the sides of the
triangle OVR is evidently equal to DV + VS, or 2DV,
or 4OV; take OV from both, and we have

$$3OV = OR + RV, \text{ and also } OR^2 - RV^2 = OV^2,$$

that is, $$\overline{OR + RV} \times \overline{OR - RV} = OV^2,$$

substituting $$3OV \times \overline{OR - RV} = OV^2,$$

hence $$OR - RV = \tfrac{1}{3}OV;$$

subtracting this last from

$$OR + RV = 3OV,$$

we have

$$2RV = 3OV - \tfrac{1}{3}OV,$$

and, therefore,

$$RV = 1\tfrac{1}{3}OV.$$

The three sides of the triangle OVR are consequently as the numbers 5, 4, 3 : and in this trapezium the bottom width is $\frac{1}{5}$th of the top, but this last relation between the top and bottom widths is not needed to the simultaneous equality of areas and borders in a trapezium and rectangle, for if the slopes of the former be $1\frac{1}{3}$ to 1 and the vertical sides of the rectangle bisect them, then, however great the bottom width RU may be, or even if it disappear and the slopes meet in a point, the condition holds, and the rectangle and trapezium have the same discharge, velocity, and hydraulic mean depth consequent on their identity of border and area.

In the woodcut, Fig. 78, the tangents from *a* and *a'* are at a slope of 2 to 1: those from *c* and *c'* are at 1 to 1. It is remarkable how small relatively the bottom width becomes as the slope of the sides becomes flatter, at $1\frac{1}{3}$ to 1, being but two-thirds of the depth and one-fifth of the top width; at 3 to 1, being one-third of the depth and a tenth of the top width; for flatter slopes than this last, the trapezium of best form may be considered practically to merge into a triangle. If the top width with the flatter slopes be considered to involve too great an expenditure in land, and that the upper part of the excavated ground be of a nature to bear a steeper slope than the lower, then tangents to the semicircle with that slope will give a figure of best discharging form for a given area, and with those conditions, economizing both land and excavation, instances of such as having been adopted in practice are given in a future page. The bottom, also, of the channel may be constructed curvilinear by adopting for it the arc of the semicircle between the points at which the side slopes touch it, and which arc, therefore, subtends an angle at the centre equal to twice the angle of inclination of the sides. In these two last modifications the hydraulic mean depth, and therefore the velocity, are evidently the same as in the simple trapezium, and the discharge diminished only as the area of either is diminished by the omitted portions of the original trapezium.

In order to compare a trapezoidal channel of best discharging figure with others having, first, the same constant border; secondly, the same constant area, and in both the inclination of the side slopes identical, we may proceed as in pp. 206, 208, in which were tabulated the results upon the hydraulic mean depth and discharge of all the different figures which a transverse section of a rectangular form may assume when the border is constant, and again when the area is constant.

Let us suppose a trapezium with side slopes of 30° with the horizon, that is, 1.732 to 1 (nearly 1¾ to 1), and a constant border of 200 units. It is evident (Fig. 79) that when we diminish the depth, the bottom width increases, and that the limiting figure is with a depth equal to zero, a bottom width equal to 200. On the ·

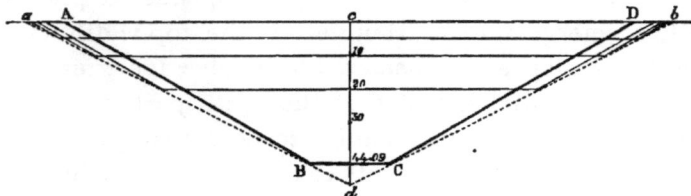

Fig. 79.

other hand, when the bottom width is equal to zero, and the depth the greatest possible, the trapezium becomes an isosceles triangle, whose base angles are equal to the angle of inclination of the sides, and consequently unlike the former limit (and both limiting figures in the rectangle of constant border) having a finite area.

TABLE *showing the Value of the Bottom Width, Border, and Hydraulic Mean Depth, &c., the Transverse Section being a Trapezium with Side Slopes 30° with Horizon, and a Constant Border equal 200 Units.*

Depth in Feet.	Bottom Width.	Area.	Hydraulic Mean Depth.	Square Root of Hydraulic Mean Depth.
0	200	0	0	0
1	196	197.7	0.99	0.99
2	192	390.9	1.96	1.39
3	188	579.6	2.90	1.70
5	180	943.3	4.72	2.17
10	160	1773.2	8.87	2.97
20	120	3092.8	15.46	3.93
25	100	3582.5	17.91	4.23
30	80	3958.8	19.79	4.45
35	60	4221.8	21.11	4.58
40	40	4371.3	21.86	4.67
41	36	4387.6	21.94	4.68
42	32	4399.3	22.00	4.69
43	28	4406.6	22.03	4.694
44	24	4409.25	22.04	4.695
44.09	23.63	4409.27	22.05	4.696
45	20	4407.3	22.04	4.694
50	0	4330.1	21.65	4.653

Taking in the next place a constant transverse area for the trapezium, as in Fig. 80, whose side slopes are

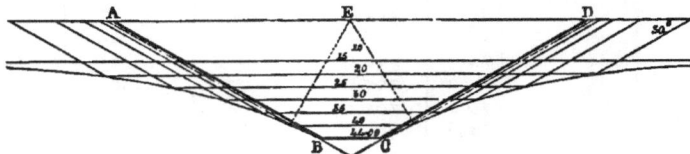

Fig. 80.

at 30° with the horizon, and supposing this area to be that which was the maximum in the above Table, 4409.27

square units: we find the mean widths by dividing this number by the several depths we assume; the quotient is the mean width of the trapezium, from which the top and bottom widths are deduced by adding and subtracting the product of the depth into the ratio of the slopes.

From this it readily appears that there is a limit to the depth, for, when the product above mentioned is equal to the mean width, the difference is zero, and the figure becomes an isosceles triangle, whose base angles are equal the angle of inclination of the sides, and area equal the constant area chosen. But the bottom width increases without limit as the depth chosen diminishes, and at a depth equal to zero becomes infinite.

TABLE *showing the Value of the Bottom Width, Border, and Hydraulic Mean Depth, &c., the Transverse Section being a Trapezium, with Side Slopes* 30° *with Horizon, and the Constant Area* 4409.27 *Square Units.*

Depth.	Bottom Width.	Border.	Hydraulic Mean Depth.	Square Root of Hydraulic Mean Depth.
0	∞	∞	0	0
1	4407.5	4411.5	0.99	0.99
2	2201.2	2209.2	1.99	1.40
3	1464.6	1476.6	2.98	1.73
5	873.2	893.2	4.93	2.22
10	423.6	463.6	9.50	3.08
20	185.8	265.8	16.60	4.07
25	133.1	233.1	18.92	4.35
30	95.0	215.0	20.50	4.53
35	65.3	205.3	21.46	4.63
40	40.9	200.9	21.95	4.685
41	36.5	200.5	21.99	4.689
42	32.2	200.2	22.02	4.692
43	28.1	200.1	22.03	4.694
44	24.0	200.0	22.04	4.695
44.09	23.6	200.0	22.05	4.696
45	20.0	200.0	22.04	4.695
50	1.6	201.6	21.88	4.67
50.455	0.	201.8	21.85	4.674

In both the Tables it is remarkable how nearly equal the numbers in the fifth column are, on each side of the maximum, even for a wide range of values assumed for the depth and bottom width, and this has an important practical bearing, for if a depth of 44 and bottom width of 24 units were found inapplicable or expensive from the great depth the excavated earth would have to be raised, we may adopt a channel having a depth of 30 and bottom width of 95 units, and as the transverse area is constant, the volume discharged will be influenced only by the alteration in the mean velocity, that is, in the value of the square root of the hydraulic mean depth, and this we perceive is only reduced from 4.695 to 4.53, which is less than $3\frac{1}{2}$ per cent.; the longitudinal inclination of the bed of the channel being the same in both cases. The land required for the wider channel would be about 12 per cent. greater than for that of best discharging form. In forming a new or an improved river channel, the excavated earth is almost always carried to spoil on each side, and not to a contiguous embankment, as in road or railway works, which makes the *depth* of the cutting of the greater importance with a view to economy.

In all cases the top width spoken of is supposed to be the level of high water of the greatest floods, and should be 3 or 4 feet below the surface of the land on each side, in order that the thorough drainage may not be injuriously affected.

116. The mean velocity of water flowing in an open channel is about 4-5ths of the maximum velocity, which is generally at the centre and upon the surface, or a little below it; and, conversely, the maximum velocity at the surface is found from the mean velocity by adding a fourth (Minard, "Cours de Construction," p. 6).

If U be the mean velocity in feet per second, and V

the observed maximum, we have, therefore, approximately—

(1) . . . . . . $U = 0.8\ V.$

Thus, if the mean velocity be 3 ft. per sec., that at the surface is $\frac{5}{4} \times 3 = 3.75$ ft. per sec.

Also, if the observed central velocity at the surface were found to be 5.2 ft., the mean velocity is 4.16 ft. per sec.

For many purposes of civil engineering this is sufficiently accurate, but MM. Darcy and Bazin have objected to it as not taking any account of the inclination of the surface, or of the hydraulic mean depth which depends on the greater or less dimensions of the transverse profile, and have given for U, the mean velocity, this formula which is adapted to measures of English feet— V being, as above, the observed central surface velocity—

(2)   . . $U = V - 25.5 \sqrt{\dfrac{h}{l}} \times H_y.$

It is stated that this was found by them to check very well with the mean velocity deduced by an independent method. At first sight this formula may seem, by comparison, to lack simplicity; but to apply the former in estimating the discharge of large rivers, it would, in the end, require the very same measurement of the transverse section as in equation (2). Dubuat has given an empiric formula,

(3) . . . . . $U = V - \sqrt{V + 0.5},$

on which have been founded Tables by many authors.

117. In the formulæ for the velocity and discharge of open channels and pipes given in this Chapter, the

direction of both the pipe and channel is supposed to be
nearly a right line: when they have quick curves, an additional resistance is occasioned, which diminishes the
discharge, or demands an increased head to give any
required charge.   This resistance is said to depend con-
jointly upon the square of the velocity of the water, upon
the number of bends, and on the square of the sine of
angle they make with the straight line of direction ; and
Mr. Beardmore has added, inversely, as the square root
of the hydraulic mean depth ; but experimenters have
not been consistent in the results obtained (D'Aubuisson, §§ 196–198).

In the cases of pipes running full, the bends may
occur in the vertical plane also; and in this case the
air is found to collect rapidly at the summit of such
bends : air-valves must, therefore, be left to free the
pipe, which may be in some cases self-acting, but are
generally worked by hand at stated times.

It was formerly thought necessary to proportion the
diameter of the main pipe in the different parts of its
course, so as to make it, at the termination, discharge the
quantity due to its diameter.   Thus, at the enlargement of
the Edinburgh Waterworks, as designed by Mr. Jardine,
the main for the first 18,300 ft. had a fall of 65 ft. ; and
the diameter, commencing at 20 inches, decreased to
18 inches : the remainder of the distance, 27,900 ft., had
a fall of 286 ft.,—nearly three times the former—and a
diameter of only 15 inches.   The discharge into the
Castle Hill distributing reservoir is only that due to
the smaller diameter, laid the whole distance with a
uniform fall.   The present and more correct practice is to
give the main a uniform diameter throughout; but at
no point in the line of pipe track must the main at any
of the vertical bends rise above the line of the average
descent on which the discharge was calculated.

118. On this main were placed, at fourteen different points,—the summits of bends in the vertical plane—cast-iron vessels to receive the compressed air as it collected. Fig. 81 shows a vertical section of one of them, 4 ft. high, and 1.5 ft. wide, with the cock for letting off the air, which must be done every three or four days. The neglect of this precaution has, in former

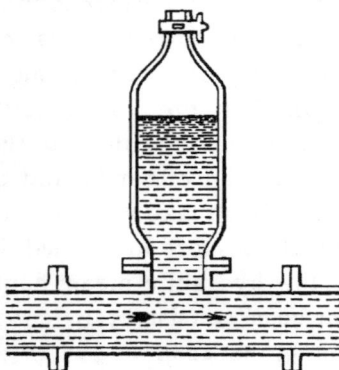

Fig. 81.

days, been the cause of great disappointment upon the first opening of waterworks. At the present time a self-acting apparatus has been adopted, which will be described in Part II., with other details which are required on the line of a main pipe track for its safe and efficient working.

39 PATERNOSTER ROW, E.C.
LONDON, *August* 1875.

GENERAL LIST OF WORKS

PUBLISHED BY

MESSRS. LONGMANS, GREEN, AND CO.

————◦∘◦————

——∘⦂∘⦂∘——

HISTORY, POLITICS, HISTORICAL MEMOIRS, &c.

Journal of the Reigns of King George the Fourth and King William the Fourth.

By the late Charles Cavendish Fulke Greville, Esq.

Edited by Henry Reeve, Esq.

Fifth Edition. 3 *vols.* 8*vo. price* 36*s.*

The Life of Napoleon III. derived from State Records, Unpublished Family Correspondence, and Personal Testimony.

By Blanchard Jerrold.

Four Vols. 8*vo. with numerous Portraits and Facsimiles.* VOLS. I. and II. *price* 18*s. each.*

*** *Vols. III. and IV. are in preparation.*

A

Recollections and Sugges-
tions, 1813–1873.
By John Earl Russell, K.G.
New Edition, revised and enlarged. 8vo. 16s.

Introductory Lectures on
Modern History delivered
in Lent Term 1842 ; *with*
the Inaugural Lecture de-
livered in December 1841.
By the late Rev. Thomas
Arnold, D.D.
8vo. price 7s. 6d.

On Parliamentary Go-
vernment in England: its
Origin, Development, and
Practical Operation.
By Alpheus Todd.
2 vols. 8vo. £1. 17s.

The Constitutional His-
tory of England since the
Accession of George III.
1760–1870.
By Sir Thomas Erskine
May, K.C.B.
Fourth Edition. 3 vols. crown 8vo. 18s.

Democracy in Europe;
a History.
By Sir Thomas Erskine
May, K.C.B.
2 vols. 8vo.   [In the press.

The History of England
from the Fall of Wolsey to
the Defeat of the Spanish
Armada.
By J. A. Froude, M.A.
CABINET EDITION, 12 vols. cr. 8vo. £3. 12s.
LIBRARY EDITION, 12 vols. 8vo. £8. 18s.

The English in Ireland
in the Eighteenth Century.
By J. A. Froude, M.A.
3 vols. 8vo. £2. 8s.

The History of England
from the Accession of
James II.
By Lord Macaulay.
STUDENT'S EDITION, 2 vols. cr. 8vo. 12s.
PEOPLE'S EDITION, 4 vols. cr. 8vo. 16s.
CABINET EDITION, 8 vols. post 8vo. 48s.
LIBRARY EDITION, 5 vols. 8vo. £4.

Critical and Historical
Essays contributed to the
Edinburgh Review.
By the Right Hon. Lord
Macaulay.
Cheap Edition, authorised and complete,
crown 8vo. 3s. 6d.
STUDENT'S EDITION, crown 8vo. 6s.
PEOPLE'S EDITION, 2 vols. crown 8vo. 8s.
CABINET EDITION, 4 vols. 24s.
LIBRARY EDITION, 3 vols. 8vo. 36s.

Lord Macaulay's Works.
Complete and uniform Li-
brary Edition.
Edited by his Sister, Lady
Trevelyan.
8 vols. 8vo. with Portrait, £5. 5s.

Lectures on the History
of England from the Ear-
liest Times to the Death of
King Edward II.
By W. Longman, F.S.A.
Maps and Illustrations. 8vo. 15s.

The History of the Life
and Times of Edward III.
By W. Longman, F.S.A.
With 9 Maps, 8 Plates, and 16 Woodcuts.
2 vols. 8vo. 28s.

History of England
under the Duke of Bucking-
ham and Charles the First,
1624–1628.
By S. Rawson Gardiner,
late Student of Ch. Ch.
2 vols. 8vo. with two Maps, 24s.

History of Civilization in
England and France, Spain
and Scotland.
By Henry Thomas Buckle.
3 vols. crown 8vo. 24s.

A Student's Manual of
the History of India from
the Earliest Period to the
Present.
By Col. Meadows Taylor,
M.R.A.S.
Second Thousand. Cr. 8vo. Maps, 7s. 6d.

Studies from Genoese
History.
By Colonel G. B. Malleson,
C.S.I. Guardian to His
Highness the Mahárájá
of Mysore.
Crown 8vo. 10s. 6d.

The Native States of
India in Subsidiary Al-
liance with the British
Government; an Historical
Sketch. With a Notice of
the Mediatized and Minor
States.
By Colonel G. B. Malleson,
C.S.I. Guardian to His
Highness the Mahárájá
of Mysore.
With 6 Coloured Maps, 8vo. price 15s.

The History of India
from the Earliest Period
to the close of Lord Dal-
housie's Administration.
By John Clark Marshman.
3 vols. crown 8vo. 22s. 6d.

Indian Polity; a View of
the System of Administra-
tion in India.
By Lieut.-Colonel George
Chesney.
Second Edition, revised, with Map. 8vo. 21s.

Waterloo Lectures; a
Study of the Campaign of
1815.
By Colonel Charles C.
Chesney, R.E.
Third Edition. 8vo. with Map, 10s. 6d.

Essays in Modern Mili-
tary Biography.
By Colonel Charles C.
Chesney, R.E.
8vo. 12s. 6d.

The Imperial and Colo-
nial Constitutions of the
Britannic Empire, includ-
ing Indian Institutions.
By Sir E. Creasy, M.A.
With 6 Maps. 8vo. 15s.

The Oxford Reformers—
John Colet, Erasmus, and
Thomas More; being a
History of their Fellow-
Work.
By Frederic Seebohm.
Second Edition. 8vo. 14s.

*The New Reformation,
a Narrative of the Old
Catholic Movement, from
1870 to the Present Time;
with an Historical Intro-
duction.*
By Theodorus.
8vo. price 12s.

*The Mythology of the
Aryan Nations.*
By Geo. W. Cox, M.A. late
Scholar of Trinity Col-
lege, Oxford.
2 vols. 8vo. 28s.

A History of Greece.
By the Rev. Geo. W. Cox,
M.A. late Scholar of
Trinity College, Oxford.
Vols. I. and II. 8vo. Maps, 36s.

*A School History of
Greece to the Death of
Alexander the Great.*
By the Rev. George W. Cox,
M.A. late Scholar of
Trinity College, Oxford;
Author of 'The Aryan
Mythology' &c.
1 vol. crown 8vo.   [In the press.

*The History of the Pelo-
ponnesian War, by Thu-
cydides.*
Translated by Richd. Craw-
ley, Fellow of Worcester
College, Oxford.
8vo. 21s.

*The Tale of the Great
Persian War, from the
Histories of Herodotus.*
By Rev. G. W. Cox, M.A.
Fcp. 8vo. 3s. 6d.

*Greek History from The-
mistocles to Alexander, in
a Series of Lives from
Plutarch.*
Revised and arranged by
A. H. Clough.
Fcp. 8vo. Woodcuts, 6s.

*General History of Rome
from the Foundation of the
City to the Fall of Au-
gustulus, B.C. 753—A.D.
476.*
By the Very Rev. C. Meri-
vale, D.D. Dean of Ely.
With 5 Maps, crown 8vo. 7s. 6d.

*History of the Romans
under the Empire.*
By Dean Merivale, D.D.
8 vols. post 8vo. 48s.

*The Fall of the Roman
Republic; a Short History
of the Last Century of the
Commonwealth.*
By Dean Merivale, D.D.
12mo. 7s. 6d.

The Sixth Oriental Monarchy; or the Geography, History, and Antiquities of Parthia. Collected and Illustrated from Ancient and Modern sources.

By Geo. Rawlinson, M.A.

With Maps and Illustrations.   8vo. 16s.

The Seventh Great Oriental Monarchy; or, a History of the Sassanians: with Notices Geographical and Antiquarian.

By Geo. Rawlinson, M.A.

8vo. with Maps and Illustrations.
[*In the press.*

Encyclopædia of Chronology, Historical and Biographical; comprising the Dates of all the Great Events of History, including Treaties, Alliances, Wars, Battles, &c. Incidents in the Lives of Eminent Men, Scientific and Geographical Discoveries, Mechanical Inventions, and Social, Domestic, and Economical Improvements.

By B. B. Woodward, B.A. and W. L. R. Cates.

8vo. 42s.

The History of Rome.
By Wilhelm Ihne.

Vols. I. and II. 8vo. 30s.   Vols. III. and IV. in preparation.

History of European Morals from Augustus to Charlemagne.
By W. E. H. Lecky, M.A.

2 vols. 8vo. 28s.

History of the Rise and Influence of the Spirit of Rationalism in Europe.
By W. E. H. Lecky, M.A.

Cabinet Edition, 2 vols. crown 8vo. 16s.

Introduction to the Science of Religion: Four Lectures delivered at the Royal Institution; with two Essays on False Analogies and the Philosophy of Mythology.
By F. Max Müller, M.A.

Crown 8vo. 10s. 6d.

The Stoics, Epicureans, and Sceptics.
Translated from the German of Dr. E. Zeller, by Oswald J. Reichel, M.A.

Crown 8vo. 14s.

Socrates and the Socratic Schools.
Translated from the German of Dr. E. Zeller, by the Rev. O. J. Reichel, M.A.

Crown 8vo. 8s. 6d.

Sketch of the History of the Church of England to the Revolution of 1688.
By T. V. Short, D.D. sometime Bishop of St. Asaph.
New Edition. Crown 8vo. 7s. 6d.

The Historical Geography of Europe.
By E. A. Freeman, D.C.L.
8vo. Maps. [In the press.

Essays on the History of the Christian Religion.
By John Earl Russell, K.G.
Fcp. 8vo. 3s. 6d.

The Student's Manual of Ancient History: containing the Political History, Geographical Position, and Social State of the Principal Nations of Antiquity.
By W. Cooke Taylor, LL.D.
Crown 8vo. 7s. 6d.

The Student's Manual of Modern History: containing the Rise and Progress of the Principal European Nations, their Political History, and the Changes in their Social Condition.
By W. Cooke Taylor, LL.D.
Crown 8vo. 7s. 6d.

The History of Philosophy, from Thales to Comte.
By George Henry Lewes.
Fourth Edition, 2 vols. 8vo. 32s.

The Crusades.
By the Rev. G. W. Cox, M.A.
Fcp. 8vo. with Map, 2s. 6d.

The Era of the Protestant Revolution.
By F. Seebohm, Author of 'The Oxford Reformers.'
With 4 Maps and 12 Diagrams. Fcp. 8vo. 2s. 6d.

The Thirty Years' War, 1618–1648.
By Samuel Rawson Gardiner.
Fcp. 8vo. with Maps, 2s. 6d.

The Houses of Lancaster and York; with the Conquest and Loss of France.
By James Gairdner.
Fcp. 8vo. with Map, 2s. 6d.

Edward the Third.
By the Rev. W. Warburton, M.A.
Fcp. 8vo. with Maps, 2s. 6d.

BIOGRAPHICAL WORKS.

Autobiography.
By *John Stuart Mill.*
8vo. 7s. 6d.

The Life and Letters of
Lord Macaulay.
By his Nephew, G. Otto
Trevelyan, M.P. for the
Hawick District of
Burghs.
2 vols. 8vo. [*In the press.*]

Admiral Sir Edward
Codrington, a Memoir of
his Life; with Selections
from his Private and
Official Correspondence.
A bridged from the larger
work, and edited by his
Daughter, Lady Bour-
chier.
With Portrait, Maps, &c. crown 8vo.
price 7s. 6d.

Life and Letters of Gil-
bert Elliot, First Earl of
Minto, from 1751 to 1806,
when his Public Life in
Europe was closed by his
Appointment to the Vice-
Royalty of India.
Edited by the Countess of
Minto.
3 vols. post 8vo. 31s. 6d.

Recollections of Past
Life.
By Sir Henry Holland,
Bart. M.D. F.R.S.
Third Edition. Post 8vo. 10s. 6d.

Isaac Casaubon, 1559-
1614.
By *Mark Pattison, Rector
of Lincoln College, Oxford.*
8vo. price 18s.

The Memoirs of Sir
John Reresby, of Thry-
bergh, Bart. M.P. for
York, &c. 1634–1689.
Written by Himself. Edit-
ed from the Original
Manuscript by James
J. Cartwright, M.A.
Cantab. of H.M. Public
Record Office.
8vo. price 21s.

Biographical and Criti-
cal Essays, reprinted from
Reviews, with Additions
and Corrections.
By *A. Hayward, Q.C.*
Second Series, 2 vols. 8vo. 28s. Third
Series, 1 vol. 8vo. 14s.

The Life of Isambard
Kingdom Brunel, Civil
Engineer.
By *I. Brunel, B.C.L.*
With Portrait, Plates, and Woodcuts.
8vo. 21s.

Lord George Bentinck;
a Political Biography.
By the Right Hon. B.
Disraeli, M.P.
New Edition. Crown 8vo. 6s.

The Life and Letters of the Rev. Sydney Smith. Edited by his Daughter, Lady Holland, and Mrs. Austin.
Crown 8vo. 2s. 6d. sewed; 3s. 6d. cloth.

Essays in Ecclesiastical Biography. By the Right Hon. Sir J. Stephen, LL.D.
Cabinet Edition. Crown 8vo. 7s. 6d.

Leaders of Public Opinion in Ireland; Swift, Flood, Grattan, O'Connell. By W. E. H. Lecky, M.A.
Crown 8vo. 7s. 6d.

Dictionary of General Biography; containing Concise Memoirs and Notices of the most Eminent Persons of all Ages and Countries. By W. L. R. Cates.
New Edition, 8vo. 25s. Supplement, 4s. 6d.

Life of the Duke of Wellington. By the Rev. G. R. Gleig, M.A.
Crown 8vo. with Portrait, 5s.

Felix Mendelssohn's Letters from Italy and Switzerland, and Letters from 1833 to 1847. Translated by Lady Wallace.
With Portrait. 2 vols. crown 8vo. 5s. each.

The Rise of Great Families; other Essays and Stories. By Sir Bernard Burke, C.B. LL.D.
Crown 8vo. 12s. 6d.

Memoirs of Sir Henry Havelock, K.C.B. By John Clark Marshman.
Crown 8vo. 3s. 6d.

Vicissitudes of Families. By Sir Bernard Burke, C.B.
2 vols. crown 8vo. 21s.

MENTAL and POLITICAL PHILOSOPHY.

Comte's System of Positive Polity, or Treatise upon Sociology. Translated from the Paris Edition of 1851–1854, and furnished with Analytical Tables of Contents. In Four Volumes, each forming in some degree an independent Treatise:—
Vol. I. *General View of Positivism and Introductory Principles.* Translated by

J. H. Bridges, M.B. *formerly Fellow of Oriel College, Oxford.* 8vo. *price* 21s.

Vol. II. *The Social Statics, or the Abstract Laws of Human Order.* Translated by Frederic Harrison, M.A. [*In Oct.*

Vol. III. *The Social Dynamics, or the General Laws of Human Progress* (*the Philosophy of History*). Translated by E. S. Beesly, M.A. *Professor of History in University College, London.* 8vo. [*In Dec.*

Vol. IV. *The Synthesis of the Future of Mankind.* Translated by Richard Congreve, M.D., *and an Appendix, containing the Author's Minor Treatises, translated by* H. D. Hutton, M.A. *Barrister-at-Law.* 8vo. [*Early in* 1876.

Order and Progress:
Part I. Thoughts on Go-
vernment; Part II. Stu-
dies of Political Crises.
By Frederic Harrison,
M.A. of Lincoln's Inn.
8vo. 14s.

Essays, Political, Social,
and Religious.
By Richd. Congreve, M.A.
8vo. 18s.

Essays, Critical and
Biographical, contributed
to the Edinburgh Review.
By Henry Rogers.
New Edition. 2 vols. crown 8vo. 12s.

Essays on some Theolo-
gical Controversies of the
Time, contributed chiefly
to the Edinburgh Review.
By Henry Rogers.
New Edition. Crown 8vo. 6s.

Democracy in America.
By Alexis de Tocqueville.
Translated by Henry
Reeve, Esq.
New Edition. 2 vols. crown 8vo. 16s.

On Representative Go-
vernment.
By John Stuart Mill.
Fourth Edition, crown 8vo. 2s.

On Liberty.
By John Stuart Mill.
Post 8vo. 7s. 6d. crown 8vo. 1s. 4d.

Principles of Political
Economy.
By John Stuart Mill.
2 vols. 8vo. 30s. or 1 vol. crown 8vo. 5s.

Essays on some Unsettled
Questions of Political Eco-
nomy.
By John Stuart Mill.
Second Edition. 8vo. 6s. 6d.

Utilitarianism.
By John Stuart Mill.
Fourth Edition. 8vo. 5s.

A System of Logic,
Ratiocinative and Induc-
tive. By John Stuart Mill.
Eighth Edition. 2 vols. 8vo. 25s.

The Subjection of Women.
By John Stuart Mill.
New Edition. Post 8vo. 5s.

Examination of Sir
William Hamilton's Phi-
losophy, and of the principal Philosophical Questions
discussed in his Writings.
By John Stuart Mill.
Fourth Edition. 8vo. 16s.

Dissertations and Dis-
cussions.
By John Stuart Mill.
Second Edition. 3 vols. 8vo. 36s. VOL. IV.
(completion) price 10s. 6d.

B

Analysis of the Pheno-mena of the Human Mind. By *James Mill. New Edition, with Notes, Illustrative and Critical.*
2 vols. 8vo. 28s.

A Systematic View of the Science of Jurispru-dence. By *Sheldon Amos, M.A.*
8vo. 18s.

A Primer of the English Constitution and Govern-ment. By *Sheldon Amos, M.A.*
Second Edition. Crown 8vo. 6s.

Principles of Economical Philosophy. By *H. D. Macleod, M.A. Barrister-at-Law.*
Second Edition, in 2 vols. Vol. I. 8vo. 15s.
Vol. II. Part I. price 12s.

The Institutes of Jus-tinian; with English In-troduction, Translation, and Notes. By *T. C. Sandars, M.A.*
Fifth Edition. 8vo. 18s.

Lord Bacon's Works, Collected and Edited by R. L. Ellis, M.A. J. Sped-ding, M.A. and D. D. Heath.
New and Cheaper Edition. 7 vols. 8vo. £3. 13s. 6d.

Letters and Life of Francis Bacon, including all his Occasional Works. Collected and edited, with a Commentary, by J. Spedding.
7 vols. 8vo. £4. 4s.

The Nicomachean Ethics of Aristotle. Newly trans-lated into English. By *R. Williams, B.A.*
8vo. 12s.

The Politics of Aristotle; Greek Text, with English Notes. By *Richard Congreve, M.A.*
New Edition, revised. 8vo. 18s.

The Ethics of Aristotle; with Essays and Notes. By *Sir A. Grant, Bart. M.A. LL.D.*
Third Edition. 2 vols. 8vo. price 32s.

Bacon's Essays, with Annotations. By *R. Whately, D.D.*
New Edition. 8vo. 10s. 6d.

Picture Logic; an At-tempt to Popularise the Science of Reasoning by the combination of Humorous Pictures with Examples of Reasoning taken from Daily Life. By *A. Swinbourne, B.A.*
With Woodcut Illustrations from Drawings by the Author. Fcp. 8vo. price 5s.

Elements of Logic.
By R. Whately, D.D.
New Edition. 8vo. 10s. 6d. cr. 8vo. 4s. 6d.

Elements of Rhetoric.
By R. Whately, D.D.
New Edition. 8vo. 10s. 6d. cr. 8vo. 4s. 6d.

An Outline of the Necessary Laws of Thought : a Treatise on Pure and Applied Logic.
By the Most Rev. W. Thomson, D.D. Archbishop of York.
Ninth Thousand. Crown 8vo. 5s. 6d.

An Introduction to Mental Philosophy, on the Inductive Method.
By J. D. Morell, LL.D.
8vo. 12s.

Elements of Psychology,
containing the Analysis of the Intellectual Powers.
By J. D. Morell, LL.D.
Post 8vo. 7s. 6d.

The Secret of Hegel :
being the Hegelian System in Origin, Principle, Form, and Matter.
By J. H. Stirling, LL.D.
2 vols. 8vo. 28s.

Sir William Hamilton ;
being the. Philósophy of Perception : an Analysis.
By J. H. Stirling, LL.D.
8vo. 5s.

Ueberweg's System of Logic, and History of Logical Doctrines.
Translated, with Notes and Appendices, by T. M. Lindsay, M.A. F.R.S.E.
8vo. 16s.

The Senses and the Intellect.
By A. Bain, LL.D. Prof. of Logic, Univ. Aberdeen.
8vo. 15s.

Mental and Moral Science ; a Compendium of Psychology and Ethics.
By A. Bain, LL.D.
Third Edition. Crown 8vo. 10s. 6d. Or separately: Part I. Mental Science, 6s. 6d. Part II. Moral Science, 4s. 6d.

The Philosophy of Necessity ; or, Natural Law as applicable to Mental, Moral, and Social Science.
By Charles Bray.
Second Edition. 8vo. 9s.

Hume's Treatise on Human Nature.
Edited, with Notes, &c. by T. H. Green, M.A. and the Rev. T. H. Grose, M.A.
2 vols. 8vo. 28s.

Hume's Essays Moral, Political, and Literary.
By the same Editors.
2 vols. 8vo. 28s.

*** The above form a complete and uniform Edition of HUME'S Philosophical Works.

MISCELLANEOUS & CRITICAL WORKS.

Miscellaneous and Posthumous Works of the late Henry Thomas Buckle. Edited, with a Biographical Notice, by Helen Taylor.
3 vols. 8vo. £2. 12s. 6d.

Short Studies on Great Subjects. By J. A. Froude, M.A. formerly Fellow of Exeter College, Oxford.
CABINET EDITION, 2 vols. crown 8vo. 12s.
LIBRARY EDITION, 2 vols. 8vc. 24s.

Lord Macaulay's Miscellaneous Writings.
LIBRARY EDITION, 2 vols. 8vo. Portrait, 21s.
PEOPLE'S EDITION, 1 vol. cr. 8vo. 4s. 6d.

Lord Macaulay's Miscellaneous Writings and Speeches.
Students' Edition.   Crown 8vo. 6s.

Speeches of the Right Hon. Lord Macaulay, corrected by Himself.
People's Edition.   Crown 8vo. 3s. 6d.

Lord Macaulay's Speeches on Parliamentary Reform in 1831 and 1832.
16mo. 1s.

Manual of English Literature, Historical and Critical. By Thomas Arnold, M.A.
New Edition.   Crown 8vo. 7s. 6d.

The Rev. Sydney Smith's Essays contributed to the Edinburgh Review.
Authorised Edition, complete in One Volume.
Crown 8vo. 2s. 6d. sewed, or 3s. 6d. cloth.

The Rev. Sydney Smith's Miscellaneous Works.
Crown 8vo. 6s.

The Wit and Wisdom of the Rev. Sydney Smith.
Crown 8vo. 3s. 6d.

The Miscellaneous Works of Thomas Arnold, D.D. Late Head Master of Rugby School and Regius Professor of Modern History in the Univ. of Oxford.
8vo. 7s. 6d.

Realities of Irish Life. By W. Steuart Trench.
Cr. 8vo. 2s. 6d. sewed, or 3s. 6d. cloth.

Lectures on the Science of Language. By F. Max Müller, M.A. &c.
Eighth Edition.   2 vols. crown 8vo. 16s.

Chips from a German Workshop; being Essays on the Science of Religion, and on Mythology, Traditions, and Customs. By F. Max Müller, M.A. &c.
3 vols. 8vo. £2.

Southey's Doctor, complete in One Volume.
Edited by Rev. J. W. Warter, B.D.
Square crown 8vo. 12s. 6d.

Families of Speech.
Four Lectures delivered at the Royal Institution.
By F. W. Farrar, D.D.
New Edition. Crown 8vo. 3s. 6d.

Chapters on Language.
By F. W. Farrar, D.D. F.R.S.
New Edition. Crown 8vo. 5s.

A Budget of Paradoxes.
By Augustus De Morgan, F.R.A.S.
Reprinted, with Author's Additions, from the Athenæum. 8vo. 15s.

Apparitions; a Narrative of Facts.
By the Rev. B. W. Savile, M.A. Author of ' The Truth of the Bible' &c.
Crown 8vo. price 4s. 6d.

Miscellaneous Writings of John Conington, M.A.
Edited by J. A. Symonds, M.A. With a Memoir by H. J. S. Smith, M.A.
2 vols. 8vo. 28s.

Recreations of a Country Parson.
By A. K. H. B.
Two Series, 3s. 6d. each.

Landscapes, Churches, and Moralities.
By A. K. H. B.
Crown 8vo. 3s. 6d.

Seaside Musings on Sundays and Weekdays.
By A. K. H. B.
Crown 8vo. 3s. 6d.

Changed Aspects of Unchanged Truths.
By A. K. H. B.
Crown 8vo. 3s. 6d.

Counsel and Comfort from a City Pulpit.
By A. K. H. B.
Crown 8vo. 3s. 6d.

Lessons of Middle Age.
By A. K. H. B.
Crown 8vo. 3s. 6d.

Leisure Hours in Town
By A. K. H. B.
Crown 8vo. 3s. 6d.

The Autumn Holidays of a Country Parson.
By A. K. H. B.
Crown 8vo. 3s. 6d.

Sunday Afternoons at the Parish Church of a Scottish University City.
By A. K. H. B.
Crown 8vo. 3s. 6d.

The Commonplace Phi-
losopher in Town and
Country.
By A. K. H. B.
Crown 8vo. 3s. 6d.

Present-Day Thoughts.
By A. K. H. B.
Crown 8vo. 3s. 6d.

Critical Essays of a
Country Parson.
By A. K. H. B.
Crown 8vo. 3s. 6d.

The Graver Thoughts of
a Country Parson.
By A. K. H. B.
Two Series, 3s. 6d. each.

DICTIONARIES and OTHER BOOKS of REFERENCE.

A Dictionary of the
English Language.
By R. G. Latham, M.A.
M.D. Founded on the
Dictionary of Dr. S.
Johnson, as edited by
the Rev. H. J. Todd,
with numerous Emenda-
tions and Additions.
4 vols. 4to. £7.

Thesaurus of English
Words and Phrases, classi-
fied and arranged so as to
facilitate the expression of
Ideas, and assist in Literary
Composition.
By P. M. Roget, M.D.
Crown 8vo. 10s. 6d.

English Synonymes.
By E. J. Whately. Edited
by Archbishop Whately.
Fifth Edition. Fcp. 8vo. 3s.

Handbook of the English
Language. For the use of
Students of the Universities
and the Higher Classes in
Schools.
By R. G. Latham, M.A.
M.D. &c. late Fellow of
King's College, Cam-
bridge ; late Professor of
English in Univ. Coll.
Lond.
The Ninth Edition. Crown 8vo. 6s.

A Practical Dictionary
of the French and English
Languages.
By Léon Contanseau, many
years French Examiner
for Military and Civil
Appointments, &c.
Post 8vo. 10s. 6d.

Contanseau's Pocket Dic-
tionary, French and Eng-
lish, abridged from the
Practical Dictionary, by
the Author.
Square 18mo. 3s. 6d.

*New Practical Diction-
ary of the German Lan-
guage; German - English
and English-German.*
By Rev. W. L. Blackley,
M.A. and Dr. C. M.
Friedländer.
Post 8vo. 7s. 6d.

*A Dictionary of Roman
and Greek Antiquities.
With 2,000 Woodcuts
from Ancient Originals,
illustrative of the Arts
and Life of the Greeks and
Romans.*
By Anthony Rich, B.A.
Third Edition. Crown 8vo. 7s. 6d.

*The Mastery of Lan-
guages; or, the Art of
Speaking Foreign Tongues
Idiomatically.*
By Thomas Prendergast.
Second Edition. 8vo. 6s.

*A Practical English Dic-
tionary.*
By John T. White, D.D.
Oxon. and T. C. Donkin,
M.A.
1 vol. post 8vo. uniform with Contanseau's
Practical French Dictionary.
[In the press.

*A Latin-English Dic-
tionary.*
By John T. White, D.D.
Oxon. and J. E. Riddle,
M.A. Oxon.
Third Edition, revised. 2 vols. 4to. 42s.

*White's College Latin-
English Dictionary;
abridged from the Parent
Work for the use of Uni-
versity Students.*
Medium 8vo. 18s.

*A Latin-English Dic-
tionary adapted for the use
of Middle-Class Schools,*
By John T. White, D.D.
Oxon.
Square fcp. 8vo. 3s.

*White's Junior Student's
Complete Latin - English
and English-Latin Dic-
tionary.*
Square 12mo. 12s.

Separately { ENGLISH-LATIN, 5s. 6d.
{ LATIN-ENGLISH, 7s. 6d.

*A Greek-English Lexi-
con.*
By H. G. Liddell, D.D.
Dean of Christchurch,
and R. Scott, D.D.
Dean of Rochester.
Sixth Edition. Crown 4to. 36s.

*A Lexicon, Greek and
English, abridged for
Schools from Liddell and
Scott's* Greek - English
Lexicon.
Fourteenth Edition. Square 12mo. 7s. 6d.

*An English-Greek Lexi-
con, containing all the Greek
Words used by Writers of
good authority.*
By C. D. Yonge, B.A.
New Edition. 4to. 21s.

C. D. Yonge's New Lexicon, English and Greek, abridged from his larger Lexicon.

Square 12mo. 8s. 6d.

M'Culloch's Dictionary, Practical, Theoretical, and Historical, of Commerce and Commercial Navigation.
Edited by H. G. Reid.

8vo. 63s.

A General Dictionary of Geography, Descriptive, Physical, Statistical, and Historical; forming a complete Gazetteer of the World.
By A. Keith Johnston, F.R.S.E.

New Edition, thoroughly revised.
[In the press.

The Public Schools Manual of Modern Geography. Forming a Companion to 'The Public Schools Atlas of Modern Geography'
By Rev. G. Butler, M.A.
[In the press.

The Public Schools Atlas of Modern Geography. In 31 Maps, exhibiting clearly the more important Physical Features of the Countries delineated.
Edited, with Introduction, by Rev. G. Butler, M.A.
Imperial quarto, 3s. 6d. sewed; 5s. cloth.

The Public Schools Atlas of Ancient Geography. Edited, with an Introduction on the Study of Ancient Geography, by the Rev. G. Butler, M.A.
Imperial Quarto.   [In the press.

ASTRONOMY and METEOROLOGY.

The Universe and the Coming Transits; Researches into and New Views respecting the Constitution of the Heavens.
By R. A. Proctor, B.A.
With 22 Charts and 22 Diagrams. 8vo. 16s.

Saturn and its System.
By R. A. Proctor, B.A.
8vo. with 14 Plates, 14s.

The Transits of Venus; A Popular Account of Past and Coming Transits, from the first observed by Horrocks A.D. 1639 to the Transit of A.D. 2012.

By R. A. Proctor, B.A.

With 20 Plates (12 Coloured) and 27 Woodcuts.   Crown 8vo. 8s. 6d.

Essays on Astronomy.
A Series of Papers on
Planets and Meteors, the
Sun and Sun-surrounding
Space, Stars and Star
Cloudlets.
By R. A. Proctor, B.A.
With 10 Plates and 24 Woodcuts. 8vo. 12s.

The Moon; her Motions,
Aspect, Scenery, and Phy-
sical Condition.
By R. A. Proctor, B.A.
*With Plates, Charts, Woodcuts, and Lunar
Photographs. Crown 8vo. 15s.*

The Sun; Ruler, Light,
Fire, and Life of the Pla-
netary System.
By R. A. Proctor, B.A.
*Second Edition. Plates and Woodcuts. Cr.
8vo. 14s.*

The Orbs Around Us; a
Series of Familiar Essays
on the Moon and Planets,
Meteors and Comets, the
Sun and Coloured Pairs of
Suns.
By R. A. Proctor, B.A.
*Second Edition, with Chart and 4 Diagrams.
Crown 8vo. 7s. 6d.*

Other Worlds than Ours;
The Plurality of Worlds
Studied under the Light
of Recent Scientific Re-
searches.
By R. A. Proctor, B.A.
*Third Edition, with 14 Illustrations. Cr.
8vo. 10s. 6d.*

Brinkley's Astronomy.
Revised and partly re-writ-
ten, with Additional Chap-
ters, and an Appendix of
Questions for Examination.
By John W. Stubbs, D.D.
and F. Brünnow, Ph.D.
With 49 Diagrams. Crown 8vo. 6s.

Outlines of Astronomy.
By Sir J. F. W. Herschel,
Bart. M.A.
*Latest Edition, with Plates and Diagrams.
Square crown 8vo. 12s.*

A New Star Atlas, for
the Library, the School, and
the Observatory, in 12 Circular Maps (with 2 Index
Plates).
By R. A. Proctor, B.A.
Crown 8vo. 5s.

Celestial Objects for Com-
mon Telescopes.
By T. W. Webb, M.A.
F.R.A.S.
*New Edition, with Map of the Moon and
Woodcuts. Crown 8vo. 7s. 6d.*

Larger Star Atlas, for the
Library, in Twelve Cir-
cular Maps, photolitho-
graphed by A. Brothers,
F.R.A.S. With 2 Index
Plates and a Letterpress
Introduction.
By R. A. Proctor, BA.
Second Edition. Small folio, 25s.

C

Dove's Law of Storms, considered in connexion with the ordinary Movements of the Atmosphere.
Translated by *R. H. Scott,* M.A.
8vo. 10s. 6d.

Air and Rain; the Be- ginnings of a Chemical Climatology.
By *R. A. Smith, F.R.S.*
8vo. 24s.

Air and its Relations to Life, 1774–1874. Being, with some Additions, a Course of Lectures deliver- ed at the Royal Institution of Great Britain in the Summer of 1874.
By *Walter Noel Hartley,* F.C.S. Demonstrator of Chemistry at King's College, London.
1 vol. small 8vo. with Illustratrations.
[*Nearly ready.*]

Magnetism and Devia- tion of the Compass. For the use of Students in Navi- gation and Science Schools.
By *J. Merrifield, LL.D.*
18mo. 1s. 6d.

Nautical Surveying, an Introduction to the Practi- cal and Theoretical Study of.
By *J. K. Laughton, M.A.*
Small 8vo. 6s.

Schellen's Spectrum Ana- lysis, in its Application to Terrestrial Substances and the Physical Constitution of the Heavenly Bodies.
Translated by *Jane* and *C. Lassell;* edited, with Notes, by *W. Huggins, LL.D. F.R.S.*
With 13 Plates and 223 Woodcuts. 8vo. 28s.

NATURAL HISTORY and PHYSICAL SCIENCE.

The Correlation of Phy- sical Forces.
By *the Hon. Sir W. R. Grove, F.R.S. &c.*
Sixth Edition, with other Contributions to Science. 8vo. 15s.

Professor Helmholtz' Popular Lectures on Scien- tific Subjects.
Translated by *E. Atkinson, F.C.S.*
With many Illustrative Wood Engravings. 8vo. 12s. 6d.

Ganot's Natural Philosophy for General Readers and Young Persons; a Course of Physics divested of Mathematical Formulæ and expressed in the language of daily life.

Translated by E. Atkinson, F.C.S.

Second Edition, with 2 Plates and 429 Woodcuts. Crown 8vo. 7s. 6d.

Ganot's Elementary Treatise on Physics, Experimental and Applied, for the use of Colleges and Schools.

Translated and edited by E. Atkinson, F.C.S.

New Edition, with a Coloured Plate and 726 Woodcuts. Post 8vo. 15s.

Weinhold's Introduction to Experimental Physics, Theoretical and Practical; including Directions for Constructing Physical Apparatus and for Making Experiments.

Translated by B. Loewy, F.R.A.S. With a Preface by G. C. Foster, F.R.S.

With 3 Coloured Plates and 404 Woodcuts. 8vo. price 31s. 6d.

Principles of Animal Mechanics.

By the Rev. S. Haughton, F.R.S.

Second Edition. 8vo. 21s.

Text-Books of Science, Mechanical and Physical, adapted for the use of Artisans and of Students in Public and other Schools. (The first Ten edited by T. M. Goodeve, M.A. Lecturer on Applied Science at the Royal School of Mines; the remainder edited by C. W. Merrifield, F.R.S. an Examiner in the Department of Public Education.)

Small 8vo. Woodcuts.

Edited by T. M. Goodeve, M.A.

Anderson's *Strength of Materials*, 3s. 6d.
Bloxam's *Metals*, 3s. 6d.
Goodeve's *Mechanics*, 3s. 6d.
——— *Mechanism*, 3s. 6d.
Griffin's *Algebra & Trigonometry*, 3s. 6d.
   Notes on the same, with Solutions, 3s. 6d.
Jenkin's *Electricity & Magnetism*, 3s. 6d.
Maxwell's *Theory of Heat*, 3s. 6d.
Merrifield's *Technical Arithmetic*, 3s. 6d.
   Key, 3s. 6d.
Miller's *Inorganic Chemistry*, 3s. 6d.
Shelley's *Workshop Appliances*, 3s. 6d.
Watson's *Plane & Solid Geometry*, 3s. 6d.

Edited by C. W. Merrifield, F.R.S.

Armstrong's *Organic Chemistry*, 3s. 6d.
Thorpe's *Quantitative Analysis*, 4s. 6d.
Thorpe *and* Muir's *Qualitative Analysis*, 3s. 6d.

Fragments of Science.

By John Tyndall, F.R.S.

New Edition, in the press.

Address delivered before the British Association assembled at Belfast.

By John Tyndall, F.R.S. President.

8th Thousand, with New Preface and the Manchester Address. 8vo. price 4s. 6d.

Heat a Mode of Motion.
By *John Tyndall, F.R.S.*
Fifth Edition, Plate and Woodcuts.
Crown 8vo. 10s. 6d.

Sound.
By *John Tyndall, F.R.S.*
Third Edition, including Recent Researches on Fog-Signalling; Portrait and Woodcuts. Crown 8vo. 10s. 6d.

Researches on Diamagnetism and Magne-Crystallic Action; including Diamagnetic Polarity.
By *John Tyndall, F.R.S.*
With 6 Plates and many Woodcuts. 8vo. 14s.

Contributions to Molecular Physics in the domain of Radiant Heat.
By *John Tyndall, F.R.S.*
With 2 Plates and 31 Woodcuts. 8vo. 16s.

Six Lectures on Light, delivered in America in 1872 *and* 1873.
By *John Tyndall, F.R.S.*
Second Edition, with Portrait, Plate, and 59 Diagrams. Crown 8vo. 7s. 6d.

Notes of a Course of Nine Lectures on Light, delivered at the Royal Institution.
By *John Tyndall, F.R.S.*
Crown 8vo. 1s. *sewed, or* 1s. 6d. *cloth.*

Notes of a Course of Seven Lectures on Electrical Phenomena and Theories, delivered at the Royal Institution.
By *John Tyndall, F.R.S.*
Crown 8vo. 1s. *sewed, or* 1s. 6d. *cloth.*

A Treatise on Magnetism, General and Terrestrial.
By *H. Lloyd, D.D. D.C.L.*
8vo. price 10s. 6d.

Elementary Treatise on the Wave-Theory of Light.
By *H. Lloyd, D.D. D.C.L.*
Third Edition. 8vo. 10s. 6d.

An Elementary Exposition of the Doctrine of Energy.
By *D. D. Heath, M.A.*
Post 8vo. 4s. 6d.

The Comparative Anatomy and Physiology of the Vertebrate Animals.
By *Richard Owen, F.R.S.*
With 1,472 *Woodcuts.* 3 *vols. 8vo.* £3. 13s. 6d.

Sir H. Holland's Fragmentary Papers on Science and other subjects.
Edited by the Rev. J. Holland.
8vo. price 14s.

Light Science for Leisure Hours; Familiar Essays on Scientific Subjects, Natural Phenomena, &c.
By *R. A. Proctor, B.A.*
First and Second Series. 2 *vols. crown 8vo.* 7s. 6d. *each.*

Kirby and Spence's Introduction to Entomology, or Elements of the Natural History of Insects.
Crown 8vo. 5s.

Strange Dwellings; a Description of the Habitations of Animals, abridged from 'Homes without Hands.'
By Rev. J. G. Wood, M.A.
With Frontispiece and 60 Woodcuts. Crown 8vo. 7s. 6d.

Homes without Hands; a Description of the Habitations of Animals, classed according to their Principle of Construction.
By Rev. J. G. Wood, M.A.
With about 140 Vignettes on Wood. 8vo. 14s.

Out of Doors; a Selection of Original Articles on Practical Natural History.
By Rev. J. G. Wood, M.A.
With 6 Illustrations from Original Designs engraved on Wood. Crown 8vo. 7s. 6d.

The Polar World: a Popular Description of Man and Nature in the Arctic and Antarctic Regions of the Globe.
By Dr. G. Hartwig.
With Chromoxylographs, Maps, and Woodcuts. 8vo. 10s. 6d.

The Sea and its Living Wonders.
By Dr. G. Hartwig.
Fourth Edition, enlarged. 8vo. with many Illustrations, 10s. 6d.

The Tropical World.
By Dr. G. Hartwig.
With about 200 Illustrations. 8vo. 10s. 6d.

The Subterranean World.
By Dr. G. Hartwig.
With Maps and Woodcuts. 8vo. 10s. 6d.

The Aerial World; a Popular Account of the Phenomena and Life of the Atmosphere.
By Dr. George Hartwig.
With Map, 8 Chromoxylographs, and 60 Woodcuts. 8vo. price 21s.

Game Preservers and Bird Preservers, or 'Which are our Friends?'
By George Francis Morant, late Captain 12th Royal Lancers & Major Cape Mounted Riflemen.
Crown 8vo. price 5s.

A Familiar History of Birds.
By E. Stanley, D.D. late Ld. Bishop of Norwich.
Fcp. 8vo. with Woodcuts, 3s. 6d.

Insects at Home; a Popular Account of British Insects, their Structure Habits, and Transformations.
By Rev. J. G. Wood, M.A.
With upwards of 700 Woodcuts. 8vo. 21s.

Insects Abroad; being a Popular Account of Foreign Insects, their Structure, Habits, and Transformations.
By Rev. J. G. Wood, M.A.
With upwards of 700 Woodcuts. 8vo. 21s.

Rocks Classified and Described.
By B. Von Cotta.
English Edition, by P. H. LAWRENCE (with English, German, and French Synonymes), revised by the Author. Post 8vo. 14s.

Heer's Primæval World of Switzerland.
Translated by W. S. Dallas, F.L.S. and edited by James Heywood, M.A. F.R.S.
2 vols. 8vo. with numerous Illustrations. [In the press.

The Origin of Civilisation, and the Primitive Condition of Man; Mental and Social Condition of Savages.
By Sir J. Lubbock, Bart. M.P. F.R.S.
Third Edition, with 25 Woodcuts. 8vo. 18s

The Native Races of the Pacific States of North America.
By Hubert Howe Bancroft.
Vol. I. Wild Tribes, their Manners and Customs; with 6 Maps. 8vo. 25s.
Vol. II. Native Races of the Pacific States. 25s.
*** To be completed early in the year 1876, in Three more Volumes—
Vol. III. Mythology and Languages of both Savage and Civilized Nations.
Vol. IV. Antiquities and Architectural Remains.
Vol. V. Aboriginal History and Migrations; Index to the Entire Work.

The Ancient Stone Implements, Weapons, and Ornaments of Great Britain.
By John Evans, F.R.S.
With 2 Plates and 476 Woodcuts. 8vo. 28s.

The Elements of Botany for Families and Schools.
Eleventh Edition, revised by Thomas Moore, F.L.S.
Fcp. 8vo. with 154 Woodcuts, 2s. 6d.

Bible Animals; a Description of every Living Creature mentioned in the Scriptures, from the Ape to the Coral.
By Rev. J. G. Wood, M.A.
With about 100 Vignettes on Wood. 8vo. 21s.

The Rose Amateur's Guide.
By Thomas Rivers.
Tenth Edition. Fcp. 8vo. 4s.

A Dictionary of Science, Literature, and Art.
Re-edited by the late W. T. Brande (the Author) and Rev. G. W. Cox, M.A.
New Edition, revised. 3 vols. medium 8vo. 63s.

On the Sensations o Tone, as a Physiological Basis for the Theory of Music.
By H. Helmholtz, Professor of Physiology in the University of Berlin.
Translated by A. J. Ellis, F.R.S.
8vo. 36s.

The History of Modern Music, a Course of Lectures delivered at the Royal Institution of Great Britain.
By *John Hullah, Professor of Vocal Music in Queen's College and Bedford College, and Organist of Charterhouse.*
New Edition, 1 vol. post 8vo. [*In the press.*

The Treasury of Botany, or Popular Dictionary of the Vegetable Kingdom; with which is incorporated a Glossary of Botanical Terms.
Edited by *J. Lindley, F.R.S. and T. Moore, F.L.S.*
With 274 Woodcuts and 20 Steel Plates. Two Parts, fcp. 8vo. 12s.

A General System of Descriptive and Analytical Botany.
Translated from the French of Le Maout and Decaisne, by Mrs. Hooker.
Edited and arranged according to the English Botanical System, by *J. D. Hooker, M.D. &c. Director of the Royal Botanic Gardens, Kew.*
With 5,500 Woodcuts. Imperial 8vo. 52s. 6d.

Loudon's Encyclopædia of Plants; comprising the Specific Character, Description, Culture, History, &c. of all the Plants found in Great Britain.
With upwards of 12,000 Woodcuts. 8vo. 42s.

Handbook of Hardy Trees, Shrubs, and Herbaceous Plants; containing Descriptions &c. of the Best Species in Cultivation; with Cultural Details, Comparative Hardiness, suitability for particular positions, &c. Based on the French Work of Decaisne and Naudin, and including the 720 Original Woodcut Illustrations.
By *W. B. Hemsley.*
Medium 8vo. 21s.

Forest Trees and Woodland Scenery, as described in Ancient and Modern Poets.
By *William Menzies, Deputy Surveyor of Windsor Forest and Parks, &c.*
In One Volume, imperial 4to. with Twenty Plates, Coloured in facsimile of the original drawings, price £5. 5s.
[*Preparing for publication.*

CHEMISTRY and PHYSIOLOGY.

Miller's Elements of Chemistry, Theoretical and Practical.

Re-edited, with Additions, by H. Macleod, F.C.S.

3 *vols.* 8*vo.* £3.

PART   I. CHEMICAL PHYSICS, 15*s.*

PART II. INORGANIC CHEMISTRY, 21*s.*

PART III. ORGANIC CHEMISTRY, *New Edition in the press.*

A Dictionary of Chemistry and the Allied Branches of other Sciences.

By Henry Watts, F.C.S. assisted by eminent Scientific and Practical Chemists.

6 *vols. medium* 8*vo.* £8. 14*s.* 6*d.*

Second Supplement to Watts's Dictionary of Chemistry, completing the Record of Discovery to the year 1873.

8*vo. price* 42*s.*

Select Methods in Chemical Analysis, chiefly Inorganic.

By Wm. Crookes, F.R.S.

With 22 *Woodcuts.   Crown* 8*vo.* 12*s.* 6*d.*

Todd and Bowman's Physiological Anatomy, and Physiology of Man.

Vol. II. with numerous Illustrations, 25*s.*

Vol. I.   New Edition by Dr. LIONEL S. BEALE, F.R.S. *Parts I. and II. in* 8*vo. price* 7*s.* 6*d. each.*

Health in the House, Twenty-five Lectures on Elementary Physiology in its Application to the Daily Wants of Man and Animals.

By Mrs. C. M. Buckton.

Crown 8*vo. Woodcuts,* 5*s.*

Outlines of Physiology, Human and Comparative.

By J. Marshall, F.R.C.S. Surgeon to the University College Hospital.

2 *vols. cr.* 8*vo. with* 122 *Woodcuts,* 32*s.*

The FINE ARTS and ILLUSTRATED EDITIONS.

Poems.

By William B. Scott.

I. Ballads and Tales.   II. Studies from Nature.   III. Sonnets &c.

Illustrated by Seventeen Etchings by L. Alma Tadema *and* William B. Scott. *Crown* 8*vo.* 15*s.*

Half-hour Lectures on the History and Practice of the Fine and Ornamental Arts.

By W. B. Scott.

Third Edition, with 50 *Woodcuts.   Crown* 8*vo.* 8*s.* 6*d.*

In Fairyland; Pictures from the Elf-World. By Richard Doyle. With a Poem by W. Allingham.

With 16 coloured Plates, containing 36 Designs. Second Edition, folio, 15s.

A Dictionary of Artists of the English School: Painters, Sculptors, Architects, Engravers, and Ornamentists; with Notices of their Lives and Works. By Samuel Redgrave.

8vo. 16s.

The New Testament, illustrated with Wood Engravings after the Early Masters, chiefly of the Italian School.

Crown 4to. 63s.

Lord Macaulay's Lays of Ancient Rome. With 90 Illustrations on Wood from Drawings by G. Scharf.

Fcp. 4to. 21s.

Miniature Edition, with Scharf's 90 Illustrations reduced in Lithography.

Imp. 16mo. 10s. 6d.

Moore's Lalla Rookh, Tenniel's Edition, with 68 Wood Engravings.

Fcp. 4to. 21s.

Moore's Irish Melodies, Maclise's Edition, with 161 Steel Plates.

Super royal 8vo. 31s. 6d.

Sacred and Legendary Art. By Mrs. Jameson.

6 vols. square crown 8vo. price £5. 15s. 6d. as follows:—

Legends of the Saints and Martyrs.

New Edition, with 19 Etchings and 187 Woodcuts. 2 vols. 31s. 6d.

Legends of the Monastic Orders.

New Edition, with 11 Etchings and 88 Woodcuts. 1 vol. 21s.

Legends of the Madonna.

New Edition, with 27 Etchings and 165 Woodcuts. 1 vol. 21s.

The History of Our Lord, with that of his Types and Precursors. Completed by Lady Eastlake.

Revised Edition, with 13 Etchings and 281 Woodcuts. 2 vols. 42s.

D

The USEFUL ARTS, MANUFACTURES, &c.

Industrial Chemistry; a Manual for Manufacturers and for Colleges or Technical Schools. Being a Translation of Professors Stohmann and Engler's German Edition of Payen's 'Précis de Chimie Industrielle,' by Dr. J. D. Barry. Edited, and supplemented with Chapters on the Chemistry of the Metals, by B. H. Paul, Ph.D.
8vo. with Plates and Woodcuts.
[*In the press.*

Gwilt's Encyclopædia of Architecture, with above 1,600 *Woodcuts.*
Fifth Edition, with Alterations and Additions, by Wyatt Papworth.
8vo. 52s. 6d.

The Three Cathedrals dedicated to St. Paul in London; their History from the Foundation of the First Building in the Sixth Century to the Proposals for the Adornment of the Present Cathedral. By W. Longman, F.S.A.
With numerous Illustrations. Square crown 8vo. 21s.

Lathes and Turning, Simple, Mechanical, and Ornamental.
By W. Henry Northcott.
With 240 Illustrations. 8vo. 18s.

Hints on Household Taste in Furniture, Upholstery, and other Details. By Charles L. Eastlake, Architect.
New Edition, with about 90 Illustrations. Square crown 8vo. 14s.

Handbook of Practical Telegraphy.
By R. S. Culley, Memb. Inst. C.E. Engineer-in-Chief of Telegraphs to the Post-Office.
Sixth Edition, Plates & Woodcuts. 8vo. 16s.

Principles of Mechanism, for the use of Students in the Universities, and for Engineering Students.
By R. Willis, M.A. F.R.S. Professor in the University of Cambridge.
Second Edition, with 374 Woodcuts. 8vo. 18s.

Perspective; or, the Art of Drawing what one Sees: for the Use of those Sketching from Nature.
By Lieut. W. H. Collins, R.E. F.R.A.S.
With 37 Woodcuts.   Crown 8vo. 5s.

Encyclopædia of Civil Engineering, Historical, Theoretical, and Practical. By E. Cresy, C.E.
With above 3,000 Woodcuts. 8vo. 42s.

*A Treatise on the Steam
Engine, in its various ap-
plications to Mines, Mills,
Steam Navigation, Rail-
ways and Agriculture.*

By J. Bourne, C.E.

With Portrait, 37 Plates, and 546 Woodcuts. 4to. 42s.

*Catechism of the Steam
Engine, in its various Ap-
plications.*

By John Bourne, C.E.

New Edition, with 89 Woodcuts. Fcp. 8vo. 6s.

*Handbook of the Steam
Engine.*

*By J. Bourne, C.E. form-
ing a* KEY *to the Author's
Catechism of the Steam
Engine.*

With 67 Woodcuts. Fcp. 8vo. 9s.

*Recent Improvements in
the Steam Engine.*

By J. Bourne, C.E.

With 124 Woodcuts. Fcp. 8vo. 6s.

*Lowndes's Engineer's
Handbook; explaining the
Principles which should
guide the Young Engineer
in the Construction of Ma-
chinery.*

Post 8vo. 5s.

*Ure's Dictionary of Arts,
Manufactures, and Mines.
Seventh Edition, re-written
and greatly enlarged by
R. Hunt, F.R.S. assisted
by numerous Contributors.*

With 2,100 Woodcuts. 3 vols. medium 8vo.
price £5. 5s.

*Practical Treatise on
Metallurgy,
Adapted from the last Ger-
man Edition of Professor
Kerl's Metallurgy by W.
Crookes, F.R.S. &c. and
E. Röhrig, Ph.D.*

3 vols. 8vo. with 625 Woodcuts. £4. 19s.

*Treatise on Mills and
Millwork.
By Sir W. Fairbairn, Bt.*

With 18 Plates and 322 Woodcuts. 2 vols.
8vo. 32s.

*Useful Information for
Engineers.
By Sir W. Fairbairn, Bt.*

With many Plates and Woodcuts. 3 vols.
crown 8vo. 31s. 6d.

*The Application of Cast
and Wrought Iron to
Building Purposes.
By Sir W. Fairbairn, Bt.*

With 6 Plates and 118 Woodcuts. 8vo. 16s.

*Practical Handbook of
Dyeing and Calico-Print-
ing.
By W. Crookes, F.R.S. &c.*

With numerous Illustrations and Specimens
of Dyed Textile Fabrics. 8vo. 42s.

Occasional Papers on Subjects connected with Civil Engineering, Gunnery, and Naval Architecture.
By Michael Scott, Memb. Inst. C.E. & of Inst. N.A.
2 vols. 8vo. with Plates, 42s.

Mitchell's Manual of Practical Assaying.
Fourth Edition, revised, with the Recent Discoveries incorporated, by W. Crookes, F.R.S.
8vo. Woodcuts, 31s. 6d.

Loudon's Encyclopædia of Gardening; comprising the Theory and Practice of Horticulture, Floriculture, Arboriculture, and Landscape Gardening.
With 1,000 Woodcuts. 8vo. 21s.

Loudon's Encyclopædia of Agriculture; comprising the Laying-out, Improvement, and Management of Landed Property, and the Cultivation and Economy of the Productions of Agriculture.
With 1,100 Woodcuts. 8vo. 21s.

RELIGIOUS and MORAL WORKS.

An Exposition of the 39 Articles, Historical and Doctrinal.
By E. H. Browne, D.D. Bishop of Winchester.
New Edition. 8vo. 16s.

Historical Lectures on the Life of Our Lord Jesus Christ.
By C. J. Ellicott, D.D.
Fifth Edition. 8vo. 12s.

An Introduction to the Theology of the Church of England, in an Exposition of the 39 Articles. By Rev. T. P. Boultbee, LL.D.
Fcp. 8vo. 6s.

Three Essays on Religion: Nature; the Utility of Religion; Theism.
By John Stuart Mill.
Second Edition. 8vo. price 10s. 6d.

Sermons Chiefly on the Interpretation of Scripture.
By the late Rev. Thomas Arnold, D.D.
8vo. price 7s. 6d.

Sermons preached in the Chapel of Rugby School; with an Address before Confirmation.
By the late Rev. Thomas Arnold, D.D.
Fcp. 8vo. price 3s. 6d.

Christian Life, its Course, its Hindrances, and its Helps; Sermons preached mostly in the Chapel of Rugby School. By the late Rev. Thomas Arnold, D.D.
8vo. 7s. 6d.

Christian Life, its Hopes, its Fears, and its Close; Sermons preached mostly in the Chapel of Rugby School. By the late Rev. Thomas Arnold, D.D.
8vo. 7s. 6d.

Synonyms of the Old Testament, their Bearing on Christian Faith and Practice. By Rev. R. B. Girdlestone.
8vo. 15s.

The Primitive and Catholic Faith in Relation to the Church of England. By the Rev. B. W. Savile, M.A. Rector of Shillingford, Exeter; Author of 'The Truth of the Bible' &c.
8vo. price 7s.

Reasons of Faith; or, the Order of the Christian Argument Developed and Explained. By Rev. G. S. Drew, M.A.
Second Edition Fcp. 8vo. 6s.

The Eclipse of Faith: or a Visit to a Religious Sceptic. By Henry Rogers.
Latest Edition. Fcp. 8vo. 5s.

Defence of the Eclipse of Faith. By Henry Rogers.
Latest Edition. Fcp. 8vo. 3s. 6d.

A Critical and Grammatical Commentary on St. Paul's Epistles. By C. J. Ellicott, D.D.
8vo. Galatians, 8s. 6d. Ephesians, 8s. 6d. Pastoral Epistles, 10s. 6d. Philippians, Colossians, & Philemon, 10s. 6d. Thessalonians, 7s. 6d.

The Life and Epistles of St. Paul. By Rev. W. J. Conybeare, M.A. and Very Rev. J. S. Howson, D.D.

LIBRARY EDITION, *with all the Original Illustrations, Maps, Landscapes on Steel, Woodcuts, &c.* 2 vols: 4to. 42s.

INTERMEDIATE EDITION, *with a Selection of Maps, Plates, and Woodcuts.* 2 vols. square crown 8vo. 21s.

STUDENT'S EDITION, *revised and condensed, with 46 Illustrations and Maps.* 1 vol. crown 8vo. 9s.

An Examination into the Doctrine and Practice of Confession. By the Rev. W. E. Jelf, B.D.
8vo. price 7s. 6d.

Fasting Communion, how Binding in England by the Canons. With the testimony of the Early Fathers. An Historical Essay.

By the Rev. H. T. Kingdon, M.A.

Second Edition. 8vo. 10s. 6d.

Evidence of the Truth of the Christian Religion derived from the Literal Fulfilment of Prophecy.

By Alexander Keith, D.D.

40th Edition, with numerous Plates. Square 8vo. 12s. 6d. or in post 8vo. with 5 Plates, 6s.

Historical and Critical Commentary on the Old Testament; with a New Translation.

By M. M. Kalisch, Ph.D.

Vol. I. Genesis, 8vo. 18s. or adapted for the General Reader, 12s. Vol. II. Exodus, 15s. or adapted for the General Reader, 12s. Vol. III. Leviticus, Part I. 15s. or adapted for the General Reader, 8s. Vol. IV. Leviticus, Part II. 15s. or adapted for the General Reader, 8s.

The History and Literature of the Israelites, according to the Old Testament and the Apocrypha.

By C. De Rothschild and A. De Rothschild.

Second Edition. 2 vols. crown 8vo. 12s. 6d. Abridged Edition, in 1 vol. fcp. 8vo. 3s. 6d.

Ewald's History of Israel.

Translated from the German by J. E. Carpenter, M.A. with Preface by R. Martineau, M.A.

5 vols. 8vo. 63s.

The Types of Genesis, briefly considered as revealing the Development of Human Nature.

By Andrew Jukes.

Third Edition. Crown 8vo. 7s. 6d.

The Second Death and the Restitution of all Things; with some Preliminary Remarks on the Nature and Inspiration of Holy Scripture. (A Letter to a Friend.)

By Andrew Jukes.

Fourth Edition. Crown 8vo. 3s. 6d.

Commentary on Epistle to the Romans.

By Rev. W. A. O'Conor.

Crown 8vo. 3s. 6d.

A Commentary on the Gospel of St. John.

By Rev. W. A. O'Conor.

Crown 8vo. 10s. 6d.

The Epistle to the Hebrews; with Analytical Introduction and Notes.

By Rev. W. A. O'Conor.

Crown 8vo. 4s. 6d.

Thoughts for the Age.
By Elizabeth M. Sewell.
New Edition. Fcp. 8vo. 3s. 6d.

Passing Thoughts on Religion.
By Elizabeth M. Sewell.
Fcp. 8vo. 3s. 6d.

Preparation for the Holy Communion; the Devotions chiefly from the works of Jeremy Taylor.
By Elizabeth M. Sewell.
32mo. 3s.

Bishop Jeremy Taylor's Entire Works; with Life by Bishop Heber.
Revised and corrected by the Rev. C. P. Eden.
10 vols. £5. 5s.

Hymns of Praise and Prayer.
Collected and edited by Rev. J. Martineau, LL.D.
Crown 8vo. 4s. 6d. 32mo. 1s. 6d.

Spiritual Songs for the Sundays and Holidays throughout the Year.
By J. S. B. Monsell, LL.D.
9th Thousand. Fcp. 8vo. 5s. 18mo. 2s.

Lyra Germanica; Hymns translated from the German by Miss C. Winkworth.
Fcp. 8vo. 5s.

Endeavours after the Christian Life; Discourses.
By Rev. J. Martineau, LL.D.
Fifth Edition. Crown 8vo. 7s. 6d.

Lectures on the Pentateuch & the Moabite Stone; with Appendices.
By J. W. Colenso, D.D. Bishop of Natal.
8vo. 12s.

Supernatural Religion; an Inquiry into the Reality of Divine Revelation.
Fifth Edition. 2 vols. 8vo. 24s.

The Pentateuch and Book of Joshua Critically Examined.
By J. W. Colenso, D.D. Bishop of Natal.
Crown 8vo. 6s.

The New Bible Commentary, by Bishops and other Clergy of the Anglican Church, critically examined by the Rt. Rev. J. W. Colenso, D.D. Bishop of Natal.
8vo. 25s.

TRAVELS, VOYAGES, &c.

Italian Alps; Sketches in the Mountains of Ticino, Lombardy, the Trentino, and Venetia.

By Douglas W. Freshfield, Editor of 'The Alpine Journal.'

Square crown 8vo. Illustrations. 15s.

Here and There in the Alps.

By the Hon. Frederica Plunket.

With Vignette-title. Post 8vo. 6s. 6d.

The Valleys of Tirol; their Traditions and Customs, and How to Visit them.

By Miss R. H. Busk.

With Frontispiece and 3 Maps. Crown 8vo. 12s. 6d.

Two Years in Fiji, a Descriptive Narrative of a Residence in the Fijian Group of Islands; with some Account of the Fortunes of Foreign Settlers and Colonists up to the time of British Annexation.

By Litton Forbes, M.D. L.R.C.P. F.R.G.S. late Medical Officer to the German Consulate, Apia, Navigator Islands.

Crown 8vo. 8s. 6d.

Eight Years in Ceylon.

By Sir Samuel W. Baker, M.A. F.R.G.S.

New Edition, with Illustrations engraved on Wood by G. Pearson. Crown 8vo. Price 7s. 6d.

The Rifle and the Hound in Ceylon.

By Sir Samuel W. Baker, M.A. F.R.G.S.

New Edition, with Illustrations engraved on Wood by G. Pearson. Crown 8vo. Price 7s. 6d.

Meeting the Sun; a Journey all round the World through Egypt, China, Japan, and California.

By William Simpson, F.R.G.S.

With Heliotypes and Woodcuts. 8vo. 24s.

The Dolomite Mountains. Excursions through Tyrol, Carinthia, Carniola, and Friuli.

By J. Gilbert and G. C. Churchill, F.R.G.S.

With Illustrations. Sq. cr. 8vo. 21s.

The Alpine Club Map of the Chain of Mont Blanc, from an actual Survey in 1863–1864.

By A. Adams-Reilly, F.R.G.S. M.A.C.

In Chromolithography, on extra stout drawing paper 10s. or mounted on canvas in a folding case, 12s. 6d.

The Alpine Club Map of the Valpelline, the Val Tournanche, and the Southern Valleys of the Chain of Monte Rosa, from actual Survey.
By A. Adams-Reilly, F.R.G.S. M.A.C.

Price 6s. on extra Stout Drawing Paper, or 7s. 6d. mounted in a Folding Case.

Untrodden Peaks and Unfrequented Valleys ; a Midsummer Ramble among the Dolomites.
By Amelia B. Edwards.

With numerous Illustrations. 8vo. 21s.

The Alpine Club Map of Switzerland, with parts of the Neighbouring Countries, on the scale of Four Miles to an Inch.
Edited by R. C. Nichols, F.S.A. F.R.G.S.

In Four Sheets, in Portfolio, price 42s. coloured, or 34s. uncoloured.

The Alpine Guide.
By John Ball, M.R.I.A. late President of the Alpine Club.

Post 8vo. with Maps and other Illustrations.

Eastern Alps.
Price 10s. 6d.

Central Alps, including all the Oberland District.
Price 7s. 6d.

Western Alps, including Mont Blanc, Monte Rosa, Zermatt, &c.
Price 6s. 6d.

Introduction on Alpine Travelling in general, and on the Geology of the Alps.
Price 1s. Either of the Three Volumes or Parts of the 'Alpine Guide' may be had with this Introduction prefixed, 1s. extra. The 'Alpine Guide' may also be had in Ten separate Parts, or districts, price 2s. 6d. each.

Guide to the Pyrenees, for the use of Mountaineers.
By Charles Packe.
Second Edition, with Maps &c. and Appendix. Crown 8vo. 7s. 6d.

How to See Norway; embodying the Experience of Six Summer Tours in that Country.
By J. R. Campbell.
With Map and 5 Woodcuts, fcp. 8vo. 5s.

Visits to Remarkable Places, and Scenes illustrative of striking Passages in English History and Poetry.
By William Howitt.
2 vols. 8vo. Woodcuts, 25s.

E

WORKS of FICTION.

Whispers from Fairy-land.

By the Rt. Hon. E. H. Knatchbull - Hugessen, M.P. Author of 'Stories for my Children,' &c.

With 9 Illustrations from Original Designs engraved on Wood by G. Pearson. Crown 8vo. price 6s.

Lady Willoughby's Diary during the Reign of Charles the First, the Protectorate, and the Restoration.

Crown 8vo. 7s. 6d.

The Folk-Lore of Rome, collected by Word of Mouth from the People.

By Miss R. H. Busk.

Crown 8vo. 12s. 6d.

Becker's Gallus; or Roman Scenes of the Time of Augustus.

Post 8vo. 7s. 6d.

Becker's Charicles : Illustrative of Private Life of the Ancient Greeks.

Post 8vo. 7s. 6d.

Tales of the Teutonic Lands.
By Rev. G. W. Cox, M.A. and E. H. Jones.
Crown 8vo. 10s. 6d.

Tales of Ancient Greece.
By the Rev. G. W. Cox, M.A.
Crown 8vo. 6s. 6d.

The Modern Novelist's Library.
Atherstone Priory, 2s. boards ; 2s. 6d. cloth.
Mlle. Mori, 2s. boards ; 2s. 6d. cloth.
The Burgomaster's Family, 2s. and 2s. 6d.
MELVILLE'S Digby Grand, 2s. and 2s. 6d.
———— Gladiators, 2s. and 2s.6d.
———— Good for Nothing, 2s. & 2s. 6d.
———— Holmby House, 2s. and 2s. 6d.
———— Interpreter, 2s. and 2s. 6d.
———— Kate Coventry, 2s. and 2s. 6d.
———— Queen's Maries, 2s. and 2s. 6d.
———— General Bounce, 2s. and 2s. 6d.
TROLLOPE'S Warden, 1s. 6d. and 2s.
———— Barchester Towers, 2s. & 2s.6d.
BRAMLEY-MOORE'S Six Sisters of the Valleys, 2s. boards ; 2s. 6d. cloth.

Novels and Tales.
By the Right Hon. Benjamin Disraeli, M.P.

Cabinet Editions, complete in Ten Volumes, crown 8vo. 6s. each, as follows :—

Lothair, 6s.	Venetia, 6s.
Coningsby, 6s.	Alroy, Ixion, &c. 6s.
Sybil, 6s.	Young Duke, &c. 6s.
Tancred, 6s.	Vivian Grey, 6s.
Henrietta Temple, 6s.	
Contarini Fleming, &c. 6s.	

Stories and Tales.
By Elizabeth M. Sewell, Author of 'The Child's First History of Rome,' 'Principles of Education,' &c.
Cabinet Edition, in Ten Volumes :—

Amy Herbert, 2s. 6d.	Ivors, 2s. 6d.
Gertrude, 2s. 6d.	Katharine Ashton, 2s. 6d.
Earl's Daughter, 2s. 6d.	Margaret Percival, 3s. 6d.
Experience of Life, 2s. 6a.	Laneton Parsonage, 3s. 6d.
Cleve Hall, 2s. 6d.	
Ursula, 3s. 6d.	

POETRY and THE DRAMA.

Ballads and Lyrics of Old France; with other Poems.
By A. Lang.
Square fcp. 8vo. 5s.

Moore's Lalla Rookh,
Tenniel's Edition, with 68 Wood Engravings.
Fcp. 4to. 21s.

Moore's Irish Melodies,
Maclise's Edition, with 161 Steel Plates.
Super-royal 8vo. 31s. 6d.

Miniature Edition of Moore's Irish Melodies, with Maclise's 161 Illustrations reduced in Lithography.
Imp. 16mo. 10s. 6d.

Milton's Lycidas and Epitaphium Damonis.
Edited, with Notes and Introduction, by C. S. Jerram, M.A.
Crown 8vo. 2s. 6d.

Lays of Ancient Rome; with Ivry and the Armada.
By the Right Hon. Lord Macaulay.
16mo. 3s. 6d.

Lord Macaulay's Lays of Ancient Rome. With 90 Illustrations on Wood from Drawings by G. Scharf.
Fcp. 4to. 21s.

Miniature Edition of Lord Macaulay's Lays of Ancient Rome, with Scharf's 90 Illustrations reduced in Lithography.
Imp. 16mo. 10s. 6d.

Horatii Opera, Library Edition, with English Notes, Marginal References and various Readings.
Edited by Rev. J. E. Yonge.
8vo. 21s.

Southey's Poetical Works with the Author's last Corrections and Additions.
Medium 8vo. with Portrait, 14s.

Poems by Jean Ingelow.
2 vols. Fcp. 8vo. 10s.
FIRST SERIES, containing 'Divided,' 'The Star's Monument,' &c. 16th Thousand. Fcp. 8vo. 5s.
SECOND SERIES, 'A Story of Doom,' 'Gladys and her Island,' &c. 5th Thousand. Fcp. 8vo. 5s.

Poems by Jean Ingelow.
First Series, with nearly 100 Woodcut Illustrations.
Fcp. 4to. 21s.

Bowdler's Family Shak-speare, cheaper Genuine Edition.

Complete in 1 vol. medium 8vo. large type, with 36 Woodcut Illustrations, 14s. or in 6 vols. fcp. 8vo. price 21s.

The Æneid of Virgil Translated into English Verse.

By J. Conington, M.A.

Crown 8vo. 9s.

RURAL SPORTS, HORSE and CATTLE MANAGEMENT, &c.

Down the Road; or, Reminiscences of a Gentle-man Coachman.
By C. T. S. Birch Rey-nardson.

Second Edition, with 12 Coloured Illustra-tions from Paintings by H. Alken. Medium 8vo. price 21s.

Blaine's Encyclopædia of Rural Sports; Complete Accounts, Historical, Prac-tical, and Descriptive, of Hunting, Shooting, Fish-ing, Racing, &c.

With above 600 Woodcuts (20 from Designs by JOHN LEECH). 8vo. 21s.

A Book on Angling: a Treatise on the Art of Angling in every branch, including full Illustrated Lists of Salmon Flies.
By Francis Francis.

Post 8vo. Portrait and Plates, 15s.

Wilcocks's Sea-Fisher-man: comprising the Chief Methods of Hook and Line Fishing, a glance at Nets, and remarks on Boats and Boating.

New Edition, with 80 Woodcuts. Post 8vo. 12s. 6d.

The Ox, his Diseases and their Treatment; with an Essay on Parturition in the Cow.
By J. R. Dobson, Memb. R.C.V.S.

Crown 8vo. with Illustrations 7s. 6d.

Youatt on the Horse. Revised and enlarged by W. Watson, M.R.C.V.S.

8vo. Woodcuts, 12s. 6d.

Youatt's Work on the Dog, revised and enlarged.

8vo. Woodcuts, 6s.

Horses and Stables.
By Colonel F. Fitzwygram, XV. the King's Hussars.

With 24 Plates of Illustrations. 8vo. 10s. 6d.

The Dog in Health and Disease.
By Stonehenge.

With 73 Wood Engravings. Square crown 8vo. 7s. 6d.

The Greyhound.
By Stonehenge.

Revised Edition, with 25 Portraits of Grey-hounds, &c. Square crown 8vo. 15s.

Stables and Stable Fittings.
By W. Miles, Esq.
Imp. 8vo. with 13 Plates, 15s.

The Horse's Foot, and how to keep it Sound.
By W. Miles, Esq.
Ninth Edition. Imp. 8vo. Woodcuts, 12s. 6d.

A Plain Treatise on Horse-shoeing.
By W. Miles, Esq.
Sixth Edition. Post 8vo. Woodcuts, 2s. 6d.

Remarks on Horses' Teeth, addressed to Purchasers.
By W. Miles, Esq.
Post 8vo. 1s. 6d.

The Fly-Fisher's Entomology.
By Alfred Ronalds.
With 20 coloured Plates. 8vo. 14s.

The Dead Shot, or Sportsman's Complete Guide.
By Marksman.
Fcp. 8vo. with Plates, 5s.

WORKS of UTILITY and GENERAL INFORMATION.

Maunder's Treasury of Knowledge and Library of Reference; comprising an English Dictionary and Grammar, Universal Gazetteer, Classical Dictionary, Chronology, Law Dictionary, Synopsis of the Peerage, Useful Tables, &c.
Fcp. 8vo. 6s.

Maunder's Biographical Treasury.
Latest Edition, reconstructed and partly rewritten, with about 1,000 additional Memoirs, by W. L. R. Cates.
Fcp. 8vo. 6s.

Maunder's Scientific and Literary Treasury; a Popular Encyclopædia of Science, Literature, and Art.
New Edition, in part rewritten, with above 1,000 new articles, by J. Y. Johnson.
Fcp. 8vo. 6s.

Maunder's Treasury of Geography, Physical, Historical, Descriptive, and Political.
Edited by W. Hughes, F.R.G.S.
With 7 Maps and 16 Plates. Fcp. 8vo. 6s.

Maunder's Historical Treasury; General Intro ductory Outlines of Uni versal History, and a Series of Separate Histories.

Revised by the Rev. G. W. Cox, M.A.

Fcp. 8vo. 6s.

Maunder's Treasury of Natural History; or Popular Dictionary of Zoology.

Revised and corrected Edition. Fcp. 8vo. with 900 Woodcuts, 6s.

The Treasury of Bible Knowledge; being a Dictionary of the Books, Persons, Places, Events, and other Matters of which mention is made in Holy Scripture.

By Rev. J. Ayre, M.A.

With Maps, 15 Plates, and numerous Woodcuts. Fcp. 8vo. 6s.

Collieries and Colliers: a Handbook of the Law and Leading Cases relating thereto.

By J. C. Fowler.

Third Edition. Fcp. 8vo. 7s. 6d.

The Theory and Practice of Banking.

By H. D. Macleod, M.A.

Second Edition. 2 vols. 8vo. 30s.

Modern Cookery for Private Families, reduced to a System of Easy Practice in a Series of carefully-tested Receipts.

By Eliza Acton.

With 8 Plates & 150 Woodcuts. Fcp. 8vo. 6s.

A Practical Treatise on Brewing; with Formulæ for Public Brewers, and Instructions for Private Families.

By W. Black.

Fifth Edition. 8vo. 10s. 6d.

Three Hundred Original Chess Problems and Studies.

By Jas. Pierce, M.A. and W. T. Pierce.

With many Diagrams. Sq. fcp. 8vo. 7s. 6d. Supplement, price 3s.

The Theory of the Modern Scientific Game of Whist.

By W. Pole, F.R.S.

Seventh Edition. Fcp. 8vo. 2s. 6d.

The Cabinet Lawyer; a Popular Digest of the Laws of England, Civil, Criminal, and Constitutional.

Twenty-fourth Edition, corrected and extended. Fcp. 8vo. 9s.

Pewtner's Comprehensive Specifier; a Guide to the Practical Specification of every kind of Building-Artificer's Work.
Edited by W. Young.

Crown 8vo. 6s.

Protection from Fire and Thieves. Including the Construction of Locks, Safes, Strong-Room, and Fireproof Buildings; Burglary, and the Means of Preventing it; Fire, its Detection, Prevention, and Extinction; &c.
By G. H. Chubb, Assoc. Inst. C.E.

With 32 Woodcuts. Cr. 8vo. 5s.

Chess Openings.
By F. W. Longman, Balliol College, Oxford.

Second Edition, revised. Fcp. 8vo. 2s. 6d.

Hints to Mothers on the Management of their Health during the Period of Pregnancy and in the Lying-in Room.
By Thomas Bull, M.D.

Fcp. 8vo. 5s.

The Maternal Management of Children in Health and Disease.
By Thomas Bull, M.D.

Fcp. 8vo. 5s.

INDEX.

www.ingramcontent.com/pod-product-compliance
Lightning Source LLC
Chambersburg PA
CBHW021947220326
41599CB00012BA/1259